Algorithms, Humans, and Interactions

Amidst the rampant use of algorithmization enabled by AI, the common theme of AI systems is the human factor. Humans play an essential role in designing, developing, and operationalizing AI systems. We have a remit to ensure those systems run transparently, perform equitably, value our privacy, and effectively fulfill human needs. This book takes an interdisciplinary approach to contribute to the ongoing development of human–AI interaction with a particular focus on the "human" dimension and provides insights to improve the design of AI that could be genuinely beneficial and effectively used in society. The readers of this book will benefit by gaining insights into various perspectives about how AI has impacted people and society and how it will do so in the future, and understanding how we can design algorithm systems that are beneficial, legitimate, usable by humans, and designed considering and respecting human values. This book provides a horizontal set of guidelines and insight into how humans can be empowered by making choices about AI designs that allow them meaningful control over AI. Designing meaningful AI experiences has garnered great attention to address responsibility gaps and mitigate them by establishing conditions that enable the proper attribution of responsibility to humans. This book helps us understand the possibilities of what AI systems can do and how they can and should be integrated into our society.

Algorithms, Humans, and Interactions

How Do Algorithms Interact with People? Designing Meaningful AI Experiences

Don Donghee Shin

Routledge
Taylor & Francis Group

NEW YORK AND LONDON

Cover Image Credit: Anton Grabolle/Better Images of AI/Human-AI collaboration/CC-BY 4.0

First edition published 2023
by Routledge
6000 Broken Sound Parkway NW, Suite 300, Boca Raton, FL 33487–2742

and by Routledge
4 Park Square, Milton Park, Abingdon, Oxon, OX14 4RN

Routledge is an imprint of Taylor & Francis Group, LLC

ISBN: 978-1-032-33358-8 (hbk)
ISBN: 978-1-032-33297-0 (pbk)
ISBN: 978-1-003-31931-3 (ebk)

DOI: 10.1201/b23083

Typeset in Minion
by Apex CoVantage, LLC

Contents

Preface

ALGORITHMS ARE PERVASIVE IN our society, and we live and work with them and AI – knowingly and unknowingly. A huge amount of data regarding our daily routines is monitored and analyzed by AI to make recommendations that manage, control, and frame our behaviors in everyday life. Algorithms have become the key organizers through which power is enacted and institutionalized in society. Whether or not algorithms promote the transformation of the economy in the direction of sustainability is defined by the way these algorithms are designed, implemented, and adopted. How algorithms are coded and trained and how we can understand their decisions are important issues in this AI era.

This book presents current theories, techniques, methods, and various sociotechnical issues related to human–AI interaction. This book is a guide to understanding the dynamics of AI in human contexts by addressing some important questions: How do we ensure AI is used for our common good? How do humans and AI interact? How is AI shaping our understanding of ourselves and our societies? How do we bridge the gap between ethical considerations and practical realities to create responsible, reliable systems? Through these questions, this book proposes a way to empower humans by enabling them to make choices about AI design, which allows them control over AI. Designing meaningful AI experiences has garnered great attention to address responsibility gaps and mitigate them by establishing conditions that enable the proper attribution of responsibility to humans (e.g., users, designers, developers, providers, and lawmakers). We should not risk losing our grip on AI, which will result in amputating human intelligence and detaching unique human value. Hence, the important task is not about how to replace humans with algorithms or machine learning but rather identifying the best way to utilize AI to enrich the human experience by empowering people. With the human-centered approach, human input is situated at the center of the

AI design and construction process. This approach maximizes the potential of both humans and algorithms, allowing them to collaborate in a way that mutually reduces algorithmic biases and improves performance and results.

This book offers an integrated analysis of the logic and social implications of algorithmic processes. Reporting on the empirical and conceptual results of scholarly research, the results of such integrated analyses are useful and constructive for understanding the relations between algorithms and humans. Thus, it presents an imperative debate about what is at stake while industry and government use AI to reshape the world. By examining the immense repercussions that algorithms will have on people and society, this book brings together various perspectives on algorithms into an integrated conceptual framework and provides a broad sociotechnical analysis, addressing the critical and ethical issues of using algorithms. Weaving together various issues of AI, this book offers compelling insights into the future of an AI-based society.

Acknowledgments

THIS WORK WAS SUPPORTED by the following institutions: the Ministry of Education in the UAE, the Department of Education in Abu Dhabi Emirate (ADEK), Zayed University, Swansea University, and the National Science Foundation of US.

I am indebted to numerous anonymous reviewers of this book for their constructive and critical feedback on its proposal. With varying capacities, many scholars and industry experts have invested their valuable time in providing me with valuable information. I am grateful to Professor Yong-Jin Park of Howard University, Professor Frank Biocca of the New Jersey Institute of Technology, and Professor Anestis Fotiadis of Zayed University.

Dr. Elliot Morsia, commissioning editor from CRC/Routledge Publishing, recognized the proposal's value. I am deeply indebted to his support and interest in this book project. I am also indebted to the anonymous reviewers of this book. I thank Professor Shuhua Zhou from the University of Missouri, Columbia, for encouraging me at an early stage of the project and for the valuable feedback on earlier versions of parts of this book.

I am grateful to these coresearchers for showing interest in my work and taking part in the journey of writing this book with me in their own unique ways. These researchers are Professor Nick Diakopoulos of Northwestern University, Shyam Sundar of Penn State University, Nickolaos Stylos of the University of Bristol, and Joon Soo Lim of Syracuse University.

Many members of personnel from Naver and Daum in South Korea have also been great sources of support for this project. I am also thankful to numerous experts in AI from Samsung for their technical advice on the parts of algorithmic systems in this book. They want to remain anonymous according to their company policies. Professor Jaemin Jung from KAIST has often offered me particular encouragement in understanding the analytic aspects of algorithms. Dr. Smith has read through the entire book to improve my English expressions, and I am grateful for his friendship and intellectual contributions during the last stages of editing this book. I am also grateful to my colleagues at the university, including Paolo Mura, John Rice, Bouziane Zaid, and Mohammad Ibahrine, for their interest, encouragement, and emotional support throughout the project.

I express my gratitude to my two daughters, Emily and Ashley, who have been a source of inspiration and fuel in pursuit of my scholarly search for truth. They have been my greatest advocates – we function as a cohesive group with unconditional help for each other and no judgment. Undoubtedly, my lifelong friend and partner, Youn-Joo, has been providing me with her endless support in everything I do, and my words cannot do enough to thank her for what her companionship means to me.

Author

D ON DONGHEE SHIN HAS been a professor of digital media at Zayed University, UAE, as well as Swansea University, UK. Over the last 25 years, he has worked at various universities in the US and South Korea, including Sungkyunwkan University, Syracuse University, Penn State University, and the University of California, Berkeley. Broadly, his research areas include human–algorithm interaction, social computing, and media analytics. His research explores the impact of algorithmic platforms in terms of ethical considerations, algorithms, human–computer interaction, and media studies. In his recent research, he has examined various mechanisms to investigate users' behavior around opaque algorithmic systems, redesign these systems to communicate opaque algorithmic processes to users, and provide them with a more informed, satisfying, and engaging interaction.

Endorsements

"Don Shin has written a must-read primer for anyone considering human-AI interaction. Read this book—and learn from one of the best."
— Kerk Kee, *Professor, Texas Tech University*

"This book is a rigorous review of the field of human-algorithm interactions. It provides a guide to understanding the dynamics of AI in human contexts by addressing meaningful questions: How do humans and AI interact? How is AI shaping our understanding of ourselves and our societies? How do we bridge the gap between ethical considerations and practical realities to make responsible, reliable systems? This book seeks to find a way to empower humans by making AI design choices that allow them meaningful control over AI. Designing meaningful AI experiences has garnered great attention to address responsibility gaps and mitigate them by establishing conditions that enable a proper attribution of responsibility for humans."
— Yujong Hwang, *Professor, DePaul University*

"As a prolific author and researcher, Don Shin offers an authoritative account of algorithms in interaction with people. Instead of speculative anecdotal stories, the readers get systematically and empirically grounded projections of the interplay between AI and human—how AI emerges as a powerful determinant of our cultural, economic and social lives. His research in this book is exemplary, inspiring, and interesting, speaking to multifaceted issues but with one underlying concern about the future role of AI."
— Yong Jin Park, *author, The Future of Digital Surveillance, Howard University and BKC Harvard University*

Introduction

THIS CHAPTER INTRODUCES THE aims and structure of this book and the individual chapters. AI systems are propagating rapidly, and it is important to see those systems from human and societal perspectives. Contemporary debate about algorithms focuses on the context of machine learning techniques or mathematical analytics of data; however, they do not focus on the societal, cultural, and ethical impact of algorithms. A better approach to design algorithms is to view them as sociotechnical systems, considering both human and nonhuman components acting together as parts of these systems. Algorithm systems are composed of one or more technological algorithms, where an algorithm reflects user knowledge, user acceptance, and social experience. The sociotechnical systems perspective, considering the dynamic relations between the technicality and humanity of AI, has implications for the design and development of AI. This book offers a roadmap for work required on the ethical and societal implications of algorithms and AI. With reference to sociotechnical system theory, the interconnectedness and mutual shaping of humans and algorithms are postulated.

SOCIOTECHNICAL PERSPECTIVE FOR ALGORITHMS

Algorithms are social constructs, as much as they are statistical programs. Just like any other technology, algorithms reflect and reproduce social dynamics, and these social dynamics are often intertwined with technical, cultural, and legal issues (Sartori & Theodorou, 2022). This is why algorithm systems are sociotechnical systems that handle human interactions with technological systems (Dolata et al., 2022; Lopez, 2021). What establishes an algorithm system as a sociotechnical system is that it is generated by or related to a system that is adopted and used by users in society. Algorithms not only present information to users but also frame social processes and practices. They are not created from a vacuum but rather are designed and developed by humans, greatly influencing humans' lives as a consequence (Benk et al., 2022). Algorithms are created within societies with the hope that they will align with existing social values and cultures. What makes AI a sociotechnical system is the presence of algorithms and technical programs that control its interaction with the AI and the set of rules in which these interactions take place (Makarius et al., 2020). The presence of human-like agents has increased in today's dynamic environment. However, human actors may not be able to understand the ways in which artificial agents learn, leading to high degrees of uncertainty and unpredictability in AI systems than in traditional systems. Another unique feature of an AI system is that its borders are fuzzy and blurry, making it difficult to ensure whether the values intended by the designers are actually embedded in the algorithms, making a phenomenon of value change of particular interest to AI systems (Yu et al., 2022). These features of AI systems raise significant questions about the designs of the AI system being produced, human roles in the production of the same, system rules and values incorporated in the process of producing AI systems, and various situations arising within and outside of the system. The sociotechnical systems perspective, considering the dynamic interactions of the technical, governance, and sociocultural or institutional elements of AI, has implications for the design and governance of AI.

Sociotechnical approaches to algorithm evaluation focus on the relationship between technology and its social environment. The sociotechnical approach to AI recognizes that an algorithmic system's results depend on reciprocal influences between social structure and technical infrastructure, as well as between instrumental and human values (Stahl et al., 2021). Decision-making does not occur through purely technical reasoning. However, a solely social approach may also be improper if it does not consider how the proposed social solutions fit the complex algorithms in decision-making processes. Algorithms may not only support people in making a decision but also mislead them, trick them into a decision, or simply be used as an excuse when a decision becomes unpopular. Thus, the social and technical components of decision-making are interwoven in different ways, requiring a broader, ecological perspective (Shin, 2022). A sociotechnical lens helps us view algorithms as not only technological artifacts but also sensitizing devices that can help us rethink entrenched premises regarding fairness, transparency, and accountability (Ziewitz,

2016). What distinguishes algorithms from conventional sociotechnical systems is the existence of mediated agents and technical protocols that manage their interactions with other components of the system (Shin & Ibarahim, 2020).

The involvement of human-like agents has increased in a dynamic, evolving online environment. The sociotechnical systems perspective, considering the dynamic integration of the technical, governance, and sociocultural or institutional components of AI, has implications for the design and governance of AI. Given the social ramifications of AI, it is important to be aware of the entanglement of algorithms with their ecology – the technological and human environments within which a particular set of instructions is interpreted and put to work. It emphasizes an algorithm's relational properties – that is, how it interacts with technologies and humans collectively. In exploring their ecology, we can elucidate key questions on fairness, transparency, accountability, and trust. Although the vast competitive benefits afforded by algorithms are obvious, specifically efficiency through impressive automation and sophisticated filtering, questions remain over the extent to which human decision-making can be processed by computers (Helberger & Diakopoulos, 2022). The prevalent practice and overreliance on algorithms have also triggered issues of potential anticompetitive behaviors, as they can easily facilitate companies in attaining and enhancing collusion without any appropriate process or agreement (Voort et al., 2022). Among the prevalent issues of concern, there is an inherent problem with algorithms, which begins at the most basic level: the human bias embedded in algorithm-based decision-making systems (Hayes et al., 2020).

Algorithm-driven reality is progressively characterized by an ecosystem of autonomous automation and sociotechnical operations. Each application and service, as well as the entire sociotechnical ecosystem, are critically dependent on other components in the ecosystem. Research demands interdisciplinary approaches to examine the interactions between humans and the algorithmic society. AI demands a multidisciplinary community. Along with other disciplines, such as economics, cognitive psychology could contribute to a holistic understanding of AI. A societal conceptualization of AI increases the understanding of AI designers to avoid possible negative impacts. The importance of the sociotechnical concept of AI lies in the research gap in current AI research. Existing research is primarily driven by technical advancements in the context of economic and social problems rather than theoretically grounded research topics. Current AI systems show weaknesses in social phenomena, such as diversity, bias, and inequality.

From a sociotechnical system perspective, this book examines the immense repercussions algorithms will have on people and society by integrating various perspectives about algorithms into a conceptual framework and provides a broad sociotechnical analysis, addressing the critical and ethical issues of algorithms. Weaving together various issues of AI, this book offers compelling insights into the future of an AI-based society. This book is a guide to understanding the dynamics of AI in human contexts by addressing meaningful questions: How do humans and AI interact? How is AI shaping our understanding of ourselves and our societies? How do we bridge the gap between ethical considerations and practical realities to create responsible, reliable systems? Through these questions, this book aims to find a way to empower humans by making AI design choices that allow them control over AI. It highlights algorithmic control and auditing as effective approaches to improving transparency and fairness around opaque AI systems. Designing meaningful AI experiences has garnered great attention to address responsibility gaps and mitigate them by establishing conditions that enable the proper attribution of responsibility to humans (e.g., users, designers, developers, business leaders, policymakers, and citizens). AI development is interdisciplinary by nature, and it will benefit from a sociotechnical approach (Asatiani et al., 2021). A sociotechnical system perspective of human-AI interactions can enrich the current approach to AI and its development by providing a basis for a discourse on the definition, implementation, and control of values in human-centered AI systems, such as the respect for user value, fairness, transparency, and explainability. Therefore, this perspective becomes especially relevant, we argue, to try to align the research methods in human-AI interactions with the expectations of real-world applications.

The summaries of the chapters are as follows:

- Chapter 1 provides broader theoretical and practical perspectives on algorithmic experience. People are increasingly experiencing the outside world through the eyes of algorithms. While the rapid implementation of algorithms has significantly improved users' experience and increased convenience, concerns like how users cognitively accept such systems or through what nature and processes they recognize and objectify the capacity for algorithmic experience form the key agenda for algorithm development. This chapter discusses how users' algorithmic experiences can be improved. By examining the acceptance model of AI in light of algorithmic experience, the chapter conceptualizes the principle of algorithmic experience as part of the analytical frame for human–AI interaction.

- Chapter 2 presents an analysis of algorithmic awareness. With the drastic surge of platform algorithmification, it is important to understand users' awareness of the use of increasingly omnipresent algorithms on the online platforms they use because those algorithms can influence users' critical decisions by filtering, mediating, and shaping their interactions. This chapter conceptualizes and theorizes the principles of algorithmic awareness and further provides how it can be practically used in industry. Thus, for algorithmic interactions, it is imperative to understand what algorithms are and take the right control of data and privacy. Algorithmic awareness means not only being able to read and understand codes but also being aware of the existence, role, and underlying repercussions of algorithms. User cognitive processes of algorithmic awareness given in this chapter offer theoretical underpinnings for human-centered algorithm systems and practical guidelines for the design of algorithms.

- Chapter 3 analyzes nudges in AI by focusing on the idea of algorithmic nudges. Algorithmic nudging via AI is becoming a popular practice. Nudge principles have been applied to algorithms. While convenient and useful, these nudges raise a series of ethical concerns about privacy, information disclosure, manipulation, and tweaking. This chapter analyzes how to ensure that algorithmic nudges are used in a positive way and whether the nudge could help achieve a sustainable way of human life. This chapter discusses the principles and dimensions of the nudging effects of AI systems on user behavior as well as how people can nudge algorithmic systems to achieve human-centered results.

- Chapter 4 questions whether algorithms are reliable and how humans trust AI. Thus, it addresses an important issue of algorithmic credibility. A fundamental issue is whether we should trust what we hear about the algorithms and their recommendations. This chapter differentiates between the credibility of claims made *about* algorithms and those that are actually made *by* algorithms. Users' sense of belief that algorithms function in a robust, constructive, and legitimate manner is critical in human–algorithm interaction. Hence, this chapter focuses on the qualities of reliable algorithms by proposing a dual process of algorithmic information processing.

- Chapter 5 analyzes one of the key concerns of AI: algorithmic bias. How can we ensure that AI systems are designed responsibly and produce effective outcomes? How can social media platforms effectively diffuse reliable information instead of amplifying misinformation? AI is as biased as humans. Bias can originate from various sources, including the design of algorithms, unintended or unanticipated use of the algorithms, or algorithmic decisions about the way data are coded, framed, filtered, or analyzed to train machine learning. Algorithm-induced biases can exert negative impacts on social interactions ranging from unintended privacy infringements to solidifying societal biases of gender, race, ethnicity, and culture. The significance of the data used in training algorithms should not be underestimated. This chapter discusses how we can avoid or minimize algorithmic bias and why humans should play a part in the datafication of algorithms.

- Chapter 6 proposes and analyzes explainable algorithms. Explainability is critical in human–AI interactions. It has become almost a general consensus that AI processes should be understandable and explainable so that AI systems can be trusted. Explanation of why algorithms make a particular decision in a particular case has been demanded by the public, society, and government. As AI faces trust issues, explainable AI is considered an alternative solution to deal with transparency problems and ensure transparency so that users can understand the internal processes of algorithmic models. This chapter discusses the effects of explainability and the way it can be incorporated into AI systems. It proposes the principle of human-interpretable explanations in AI by discussing the dimensions and effects of interpretability and understandability on user attitudes and heuristics.

- Chapter 7 analyzes algorithmic journalism through a case study of the same in South Korea. Algorithms have transformed journalism in terms of news production, newsroom structure, and overall journalistic activities. Journalists around the world are trying to figure out how to make use of algorithms to improve user experience and journalism services. Using Naver's AI-based recommendation system as a case study, this chapter discusses the methods and services of algorithmic journalism, showing how an algorithm functions in news services; how it is used, processed, and understood in different journalistic contexts via different tools and approaches; and how it is communicated to users. For the sustainability of algorithmic journalism, algorithmic designers should understand journalistic values and integrate them into the construction of algorithms. Algorithmic journalism involves serious ethical considerations regarding fairness, transparency, accountability, and explainability.

- Chapter 8 proposes, defines, and conceptualizes the notion of human-centered AI based on the preceding chapters. The idea of meaningful human control has been proposed in human-centered AI, emphasizing that humans should have an ultimate grip on AI and algorithms. Human-centered AI is continuously advancing user interaction while offering effective interaction between AI and humans. A human-centered AI framework can lead to fairer, more transparent, more accountable, and more explainable AI, supporting human values, preserving human rights, and promoting user control to steer future AI in the right direction. The chapter discusses how AI should be designed and developed in a way that is human-centered and meaningfully controllable to contribute to fairer and more transparent design to forge key positive effects with the clear accountability of AI. It proposes the idea of meaningful human control as a key underpinning value in human-centered AI. Meaningful human control will play a key role in the conceptualization of a new paradigm for human–AI interaction as well as in the development of extended AI by providing theoretical underpinnings of ethical considerations and paving the practical way for human control over algorithms in AI.

The chapters are structured to consider AIs through the lens of sociotechnical systems and to embrace the complexity of all possible interactions that humans may have with AI systems. The common thread in these eight chapters is the human factor. Humans are essential in the design, development, and operation of AI systems. We have a remit to ensure those systems run transparently, perform equitably, value our privacy, and effectively fulfill human needs. AI's long-term sustainable success hinges upon our recognition that humans are critical in its design, performance, and use.

Algorithms are pervasive today. Knowingly and unknowingly, we live with and work with algorithms. Machine learning organizes thought and action. Algorithms have become the key organizers through which power is enacted in society (Dwivedi et al., 2021). Whether or not algorithms promote the transformation of the economy in the direction of sustainability will be defined by the way these algorithms are implemented (Shin et al., 2019). The way algorithms are coded and trained and how we can understand their decisions are important issues in this AI era. A huge amount of data regarding our daily routines is monitored and analyzed to make recommendations that manage, control, and lead our behaviors in everyday life (Danaher et al., 2017). The contributions in this book offer an integrated analysis of the logic and social implications of algorithmic processes, which pave the way for more responsible, trustworthy AI developments in societies, whereby humans can grip better control of even inscrutable AI. Reporting from the cutting edge of scholarly research, the results are useful and constructive for understanding the relationship between algorithms and humans. This is an imperative debate regarding what is at stake, as industry and government use AI to reshape the world.

This book concludes with a call for a diverse approach within the AI community and richer knowledge about narratives, as they help in better addressing future AI developments, public discourse, and governance. It is critical to bring together diverse perspectives and frames of approaches to correctly reflect AI in both technical and social conditions. AI practice is interdisciplinary by nature and can benefit from a sociotechnical approach.

REFERENCES

Asatiani, P. M., Nagbøl, P. R., Penttinen, E., Rinta-Kahila, T., & Salovaara, A. (2021). Sociotechnical envelopment of artificial intelligence: An approach to organizational deployment of inscrutable artificial intelligence systems. *Journal of the Association for Information Systems, 22*(2), 325–352. doi:10.17705/1jais.00664

Danaher, J., Hogan, M. J., Noone, C., Kennedy, R., Behan, A., De Paor, A., Felzmann, H., Haklay, M., Khoo, S.-M., Morison, J., Murphy, M. H., O'Brolchain, N., Schafer, B., & Shankar, K. (2017). Algorithmic governance: Developing a research agenda through the power of collective intelligence. *Big Data & Society, 4*(2). https://doi.org/10.1177/2053951717726554

Dolata, M., Feuerriegel, S., & Schwabe, G. (2022). A sociotechnical view of algorithmic fairness. *Information Systems Journal, 32*(4), 754–818.

Dwivedi, Y. K., Hughes, L., Ismagilova, E., Aarts, G., Coombs, C., Crick, T., . . ., Eirug, A. (2021). Artificial intelligence (AI): Multidisciplinary perspectives on emerging challenges, opportunities, and agenda for research, practice and policy. *International Journal of Information Management, 57* (2021), Article 101994. doi:10.1016/j.ijinfomgt.2019.08.002

Hayes, P., van de Poel, I., & Steen, M. (2020). Algorithms and values in justice and security. *AI & Society, 35,* 533–555. https://doi.org/10.1007/s00146-019-00932-9

Helberger, N., & Diakopoulos, N. (2022). The European AI Act and how it matters for research into AI in media and journalism. *Digital Journalism.* doi:1 0.1080/21670811.2022.2082505

Lopez, P. (2021). Bias does not equal bias: A socio-technical typology of bias in data-based algorithmic systems. *Internet Policy Review, 10*(4). https://doi. org/10.14763/2021.4.1598

Makarius, E., Mukherjee, D., Fox, J., & Fox, A. (2020). Rising with the machines: A socio-technical framework for bringing artificial intelligence into the organization. *Journal of Business Research, 120,* 262–273. doi:10.1016/j. jbusres.2020. 07.045

Shin, D. (2022). How do people judge the credibility of algorithmic sources? *AI and Society, 37,* 81–96. https://doi.org/10.1007/s00146-021-01158-4

Shin, D., Fotiadis, A., & Yu, H. (2019). Prospectus and limitations of algorithmic governance: An ecological evaluation of algorithmic trends. *Digital Policy, Regulation and Governance, 24*(4), 369–383. http://doi.org/10.1108/DPRG-03-2019-0017

Shin, D., & Ibarahim, M. (2020). The socio-technical assemblages of blockchain system: How blockchains are framed and how the framing reflects societal contexts. *Digital Policy, Regulation and Governance, 22*(3), 245–263. http://doi:10.1108/DPRG-11-2019-0095

Stahl, B., Andreou, A., Brey, P., Hatzakis, T., Kirichenko, A., Macnish, K., & Wright, D. (2021). Artificial intelligence for human flourishing – Beyond principles for machine learning, *Journal of Business Research, 124,* 374–388.

Ziewitz, M. (2016). Governing algorithms: Myth, mess, and methods. *Science, Technology, & Human Values, 41*(1), 3–16. https://doi. org/10.1177/0162243915608948

CHAPTER 1

Algorithmic Experience

A LGORITHMS ARE INITIATED BY human cognition, made strong by human data, and useful only when they positively influence the user experience. AI driven by machine learning algorithms is rapidly transforming human experience. People are increasingly experiencing the outside world through the eyes of algorithms. While the rapid implementation of algorithms has significantly improved users' experience and increased convenience, it is unclear how users cognitively accept such systems or through what nature and processes they recognize and objectify the capacity for algorithmic experience. Algorithmic experience is not only formed by users but in reference to a set of algorithmic performances and other parameters such as data, interface, and datafication that, at the same time, help to structure and learn from the reflexive practice. The question of how we can improve users' algorithmic experiences is becoming a key agenda for industry and academia. By discussing the algorithmic experience to examine the acceptance model of AI, we conceptualize the principle of algorithmic experience as part of the analytic frame for human–AI interaction. Algorithmic experience is essentially based on the interpretation of transparency, fairness, and accountability, along with other conventional factors of user experience, including usefulness and ease of use.

1.1 INTERACTING WITH ALGORITHMS: HOW PEOPLE PERCEIVE, COGNIZE, AND ENGAGE WITH ALGORITHMS

AI has progressively become an essential and significant component of people's everyday activities through a wide variety of applications, such as chatbot interactions, online shopping, content recommendations, personalized content aggregation services, and autonomous systems (Dwivedi et al., 2021). Algorithms curate our digital bubble with things we like by framing, prioritizing, grouping, assorting, and screening data (Bonini & Gandini, 2019). Through this curation, algorithms exert the power to shape not only users' experiences but also the formation of AI as a whole (Beer, 2017). Although the fast provenance of algorithm services has the potential to critically advance users' experience and increase satisfaction, it is still an open question how users cognitively accept such algorithmic systems (Wilson, 2017); what mechanisms lead to user satisfaction, acceptance, and trust in these algorithmic systems; or how we can improve users' AI experiences and algorithmic interactions.

Despite the holistic impact of AI on reality, it remains to be defined not only by how people experience or enjoy AI but also in what way their experience with algorithms may be furthered by automated processes (Alvarado & Waern, 2018). Algorithm services are supposed and designed to advance the user experience, but how users improve their experience through algorithms also remains unanswered. Thus far, our knowledge of how people perceive and experience algorithms is limited, although a few initial research has examined how users experience algorithm-generated news recommendation (Shin & Park, 2019) and how users perceive algorithmic recommendations that are configured and processed by algorithms (Elliott, 2021). These questions are nicely answered within the user experience (UX) frame, which seeks to improve an experience or design for a specific new experience in interaction design processes. If used correctly, the UX frame can aid the development of algorithms as a human-centered way of addressing how we can design a better UX, given the affordances that algorithms can offer (Ettlinger, 2018). By nature, AI has been designed in an inside-out manner, meaning that algorithms are based on what programmers are technically capable of doing instead of what value could be delivered to end users (Knijnenburg et al., 2012). UX frames can turn this inside-out frame into an outside-in mechanism by highlighting the affordances of AI in design and tuning in their developmental approach. The UX frame is an effective tool for humanizing AI. Given these human-centered focuses, a growing body of research has begun to scrutinize algorithmic experience. For example, Shin et al. (2020) researched algorithmic experience from an overarching perspective of user interaction with intelligent systems. It is further argued by Shin (2020) that algorithmic experience includes the ability for user control over algorithmic decision-making, transparently increasing awareness of how the system works, and consciously managing algorithmic bias and negative influence. The concept of meaningful user control over AI has increasingly been proposed as a key component of the algorithmic experience, which can also address the issues of fair, transparent, and accountable AI.

Alvarado and Waern (2018) proposed the idea of algorithmic experience as a logical frame for analyzing behavior and interaction with algorithms by illustrating how social media users feel about algorithmic computerization and how user awareness influences their interactions with the algorithms. Shin et al. (2020) further developed the term by arguing that algorithmic experience can increase users' awareness of algorithmic influence and foreground algorithmic behavior. Both studies proposed several important criteria of algorithmic experience for algorithmic systems: profiling transparency, evaluating fairness, user control, judging accountability, and profiling management. While seemingly similar, algorithmic experience is different from widely known UX. Some view algorithmic experience as a subset of UX, while others consider algorithmic experience to be a different dimension from UX. The difference can be explained by saying that algorithmic experience is a personalized user experience involving algorithms. While UX can be seen as an optimizing problem in which the goal is to improve the utility and benefits for an individual user, algorithmic experience is much more focused on reducing the harmful effects of AI and machine learning. For example, Netflix has a sophisticated UX design but a low algorithmic experience, meaning the system gives users useful and convenient personalization, but users have no idea how the results are generated. While the system may accommodate a particular goal of the users (automatizing recommended content), thus presenting a great overall user experience, the mechanisms of the inside working system may not be transparent and/or visible. A good algorithmic experience enables users to look inside the working system, understand the data collection process, and clarify where the accountability is. Thus, algorithmic experience inherently involves the fairness, transparency, and accountability of AI. For example, if users cannot understand how the codes of algorithms are configured or the hidden strategies of platforms, users cannot use AI and accept the result with full confidence.

These issues of fairness, transparency, and accountability serve as mechanisms that allow users to manage, control, and corroborate the algorithm (Diakopoolus & Koliska, 2017). When users are capable of managing how algorithms treat them and configure them, they feel empowered (Lee et al.,

2019). According to Shin (2020), the majority of users want the option to control algorithm filters, such as newsfeeds and message posts. People wish they could filter the content themselves by controlling and adjusting LinkedIn's "People You May Like" or TikTok's "For You" features, for example. In this regard, research has shown that the engagement of users in the process of designing AI systems can improve users' perceived transparency, fairness, and accountability, which then leads to establishing user trust in algorithms and AI. The mechanisms of fairness, transparency, and accountability increase users' algorithm awareness and understanding of algorithmic decision-making, leading to more informed and engaged user interactions (Shin et al., 2022a). These issues continue to become more prevalent as subjects for algorithmic experience, the algorithm acceptance model, and algorithmic design and development. While these issues become de facto standards for designing responsible and fair AI systems, specific design methods for these systems are lacking (Elliot, 2021).

1.2 THE FUNCTIONS AND ACCEPTANCE OF ALGORITHMS

The growth of advanced data analytics has led to an increase in the adoption of machine learning across a range of sectors. In the area of marketing, algorithms are deployed based on user data to predict user behaviors regarding how they shop, buy, and review products or brands. With the power of algorithms, services are being increasingly driven by data analytics and have become far more fitted and personalized based on specific user experiences. For instance, Spotify utilizes an algorithm that uses a listener's previous music habits to predict what music they like. Now about 40% of music choice on Spotify is made through recommendations suggested by Spotify's machine learning and algorithmic filtering (Bonini & Gandini, 2019). Naver, South Korea's homegrown platform, uses a mixture of personalization algorithms to rank the content based on its engagement scores to specific viewers, which helps determine what is recommended to the viewers in their accounts. TikTok's "For You" feature provides a continuous stream of content shaped by users' viewing and creation data, and the "Following" service offers a familiar and semi-chronological feed of videos from accounts that users choose to follow. Netflix algorithms trace user data in terms of average time spent, movies viewed, click-through rates, reviews, and hundreds of feedback data points (Lemos & Pastor, 2020). LinkedIn, Amazon, and YouTube leverage recommender systems to optimize users' experience when finding useful and relevant information, content, and products, forging a meaningful user experience while increasing revenue. Pandora and Apple Music take the music songs people have listened to in the past to predict what people will enjoy now and in the future. These platforms have more data, sophisticated software to analyze and make sense of user data, and thus more power than other non-platform providers. These platforms use algorithms that operate by filtering and predicting user interests and preferences for items through various machine-learning techniques, including content-based filtering and collaborative filtering mechanisms. Amazon uses content-based filtering, which selects information based on semantic content. Netflix uses collaborative filtering, combining the opinions of other customers, to make a forecast for a target user. Naver uses a blend of content-based filtering and collaborative filtering models. The combination of machine learning and algorithms brings the formation of algorithm commerce: tracing buyer behavior every second and delivering the products the buyer is most likely to purchase.

As users have deeper interaction with AI, algorithmic experience has recently garnered great attention, generating new insights and experiences through the use of algorithms (Duan et al., 2019). In particular, the role of algorithms in user experience has garnered the attention of researchers in the field of human–computer interaction. Alvarado and Waern (2018) defined algorithmic experience as the channel through which users experience systems and interfaces that are influenced by algorithmic actions. Among the public, however, there is low awareness, despite people's increased exposure to algorithms in recent years. Some users are aware of the exposure, but the algorithmic experience is not always precise or satisfying. The technical problem may be partially due to the flawed design of algorithms, which makes the algorithmic experience undesirable. A user-centered approach toward eliciting desirable algorithmic experience qualities is not straightforward since the general awareness of algorithmic influence is low among users. Thus, serious challenges exist concerning how to design and develop algorithmic interfaces. Algorithmic experience can be used as an analytical framework for making the interaction with and experience of algorithms explicit (Rossiter & Zehile, 2015). However, algorithms can exhibit unexpected negative behavior, such as unfair processes, biased results, and incorrect recommendations (Shin & Park, 2019). Another challenge to designing algorithms is that humans, unlike dehumanized robots, have emotions and feelings that are not just the totality of their behaviors. Humans do not always behave rationally or even predictably irrationally. Thus, it is essential to view algorithms not only as working tools inside a system but also as technical features deserving design attention from a user-based viewpoint. From a system perspective, the term algorithmic experience includes any possible codes or effects that could be related to the algorithmic experience or service (Courtois & Timmermans, 2018). Just like the main principle of UX, focusing on algorithmic experience proves that users can and do influence algorithms in their inaction with them, even when they do not fully understand the technicalities. This argument sheds light on the active role of users in algorithmic power from a network-down-toward-user perspective (Shin, 2020). Focusing on outcomes and the methodological shift toward a user-based approach implies the significance of the user perspective in algorithm research.

In line with the importance of the user viewpoint, it may be useful to examine what users do with algorithms, as opposed to what algorithms do to users. We thus aim to conceptualize algorithmic experience and propose an algorithm acceptance model based on algorithmic experience. Industry is becoming increasingly interested in understanding how to improve algorithmic experience and how to promote a positive UX of the use of algorithm artifacts. This algorithmic experience includes not only usability but also other cognitive, sociocognitive, and affective dimensions of users' experience in their interaction with algorithms, such as users' trust in the services or providers and users' perceptions of fairness and transparency (Lee, 2018). The key element of an algorithm is the users' trust in it; thus, it is worthwhile to examine how trust is processed in the course of algorithm adoption. As algorithms have a greater influence on people's choices than advice from humans (Beer, 2017), it can be reasonably thought that trust plays a certain role in the algorithm experience. Research on algorithms and the cognate experience is significant and timely since we are approaching the algorithm era, where almost everything is based on algorithmic functions (Shin, 2019). Thus, the success of algorithmic systems is largely positively influenced by the extent to which they promote quality experiences among their users. This discussion contributes to developing an understanding of the interactive dynamics of users, algorithms, and experience. The discussion contributes not only to the scholarly literature on human–algorithm interaction but also to practical implications concerning the design of human–algorithm interfaces.

The algorithm acceptance model is meaningful as it advances the current technology adoption model by recognizing contextual issues and the essential relationships among them (Shin et al., 2020). Algorithm technologies are increasingly characterized as an ecosystem of complex sociotechnical matters (Alvarado & Waern, 2018). By conceptualizing and developing a scale to measure algorithmic experience components, this work contributes to the research body on how to warrant such illusive and equivocal questions in algorithms and how to design algorithm systems that are user-centered and socially responsive in an algorithm era. The algorithmic experience discussed in this chapter advances the current literature on UX by highlighting algorithm motivation and behavior. Although a technology acceptance model and UX are still useful, the model is designed for general technology and the experience for general system experience. Algorithms differ from conventional technologies in many ways and require algorithm-specific factors in the model. The core of the algorithm acceptance model is on context or environment, which is better suited to algorithmic contexts. As innovative algorithm services rapidly develop, the traditional technology-based frame or conventional user framework must be modified to reflect ever-changing computing paradigms. A grasp of how people recognize algorithm functionality, how their attitudes are shaped, how behavioral intentions are performed, what cognitive views are held, and what results are derived from the cognitive process is important. The results, particularly regarding the user-based approach and perception-based quality scale, will enable future endeavors to take significant steps toward developing a human-centered algorithm framework.

1.3 HEURISTIC–SYSTEMATIC PROCESS

How do algorithmic recommendations affect human decisions, and how do users use such recommendations? A series of Shin's studies (2020, 2021, 2022) show that algorithmic curations and filtering influence user decision-making by affecting user belief about both service features (heuristics) and how to value those services (systematic evaluation). While interesting and relevant theoretically, the findings also raise practical concerns concerning the side effects of algorithmic curation on modifying and manipulating individual users' preferences as well as the diffusion of algorithmic effects to broader contexts. These concerns are increasing as AIs and algorithms are becoming increasingly common, sophisticated, and targeted (Ziewitz, 2016). With the existence of algorithms in everyday use, people have developed a sense of algorithmic experience, but they still lack the confidence to explore or comprehend them. How users' heuristics can enhance algorithmic experience has been researched as a key topic in

AI communities. Relevant research has revealed that users have limited mental models of algorithms (Sundar et al., 2022). In this light, the cognitive aspect of algorithmic experience has been researched to address how users come to understand algorithms and how their ordinary encounters with algorithms form these sensemaking processes (Shin et al., 2022b). The cognitive aspect of algorithms is based on the assumption that human intelligence can be exactly described, that an algorithm can be designed to simulate it. Research on algorithmic experience has also drawn on perceived transparency and fairness to explain users' sensemaking processes of algorithms (Shin, 2020; Shin et al., 2020). Sensemaking works to address the questions of the black box nature of algorithms (Rudin & Radin, 2019), which raises questions regarding how credible the outputs are, how fair the internal process is in making decisions, and to what extent we can trust the algorithmic outcome (Lee et al., 2019). Issues such as how we can ensure that algorithms work in fair and transparent ways (Diakopoulos, 2016), how we can design AI to have more responsibility (i.e., the algorithm should be held accountable for the outputs/recommendations; Diakopoulos, 2016), and how we can design AI systems to work best for or with people remain unclarified and controversial.

Recent research has adopted a heuristic systematic model to investigate users' sensemaking processes of algorithm properties and the perceptual experience properties they use when using and accepting AI systems (Shin et al., 2022b). The heuristic systematic model is a dual-process theory that provides a conceptual frame for understanding how users process information, establish trust judgments and make their decisions (Chaiken, 1980). The model explains users' social judgments, such as attitudes, impressions, and beliefs, and proposes two parallel ways of processing information – heuristic processing and systematic processing – which incorporate "the objective properties of the stimuli we think about and the properties that we bring into the perceptual experience" (Chaiken, 1980, p. 24). In heuristic processing, users consider available informational cues and conduct preliminary evaluations based on these limited cues. In systematic processing, users use all relevant information systematically, intensively consider that information, and form rational evaluations based on this intensive consideration (Shin, 2020). Systematic processing requires the information recipient to spend more cognitive effort judging the available information than heuristic processing does. These processes normally occur concurrently and can interact with each other during an individual's decision-making process. The model is extensively used to describe an individual's information processing and decision-making in the context of technology adoption.

The heuristic systematic model can be used to explain the algorithm acceptance model and, thus, the algorithmic experience. Shin (2020, 2021, 2022) showed that algorithm adoption involves several factors from both heuristic and systematic processes. Fairness, transparency, and accountability (and ethical factors or explainability) derive from the heuristic process, and accuracy, personalization, and predictability are adopted from the systematic process. Shin et al. (2021) argued that users see the issues of fairness, transparency, and accountability in a heuristic way since these issues are inherently hypothetical, and normally there is no solid objective way to judge them. For these issues, people rely on heuristics since evaluating fairness and transparency at the end-user level is technically unfeasible, and judging accountability as an individual user is realistically impossible (Sloan & Warner, 2018). Only what users can perceive can they also judge (Gansser & Reich, 2021). However, users involve deliberative and calculated evaluations of algorithmic performance, such as whether algorithmic results are accurate (accuracy), how the results are personally tailored (personalization), and to what extent the results are predictable and prescriptive (prescriptive predictability).

Fairness in AI indicates the principle of impartial and inclusive representation and treatment to achieve desired outcomes within the context of use (Ismagilova et al., 2022). The issue of fairness is based on the implication that algorithms do not always perform in a fair, reasonable, and non-discriminatory manner (Diakopoulos, 2016). Users expect AI to promote fair and meaningful interaction to maintain a more efficient datafication, avoiding artificially forging Likes, Shares, or Followers. The fairness of an algorithm can be objectively evaluated based on its accuracy (i.e., the percentage of correct outputs), precision (i.e., the capacity to produce precise outputs), and recall (i.e., the capability to find relevant outputs). With these criteria, evaluating fairness is not easy, as they are developed by humans who are particularly vulnerable to biases. Transparency refers to how AI makes observable and auditable what the algorithm knows about users and explains why the algorithm produces results based on that profiling, which in turn improves the algorithmic experience. As with fairness, there is no easy way for users to look into the inside of the coding for AI and algorithms. The notion of algorithmic accountability means that AI should be held responsible for the results of its algorithms (Diakopoulos, 2018), including the goals, structures, and functions. Numerous cases exemplify that algorithms could go wrong in different situations, such as facial recognition AI labeling people wrongly and autonomous vehicles guiding drivers in the wrong direction. Such mistakes may lead to serious consequences, but there is no way for end users to evaluate this accountability in their experiential dimensions. Numerous studies (e.g., Shin & Park, 2019) have examined how to conceptualize the issues of fairness, transparency, and accountability and the ways to reflect them in user interface design in AI contexts. Diakopoulos (2016) examined the effects of users' judgments of fairness and transparency on user trust and adoption decisions. Other research has also confirmed that users involve a heuristic-systematic process when evaluating the ethical issues of algorithms (Shin, 2020).

1.4 THE ALGORITHM ACCEPTANCE MODEL: HOW PEOPLE ACCEPT ALGORITHMS

An understanding of what influences user adoption of AI systems is critical for achieving the maximum potential of algorithms. To understand what influences users' adoption of AI systems, it may be logical to consider the use of already established and validated acceptance models, such as the technology acceptance model (TAM; Davis, 1989), the theory of reasoned action (Ajzen & Fishbein, 1980), and the diffusion of innovation and theory (Rogers, 2003). However, algorithm adoption may differ from other technologies and is perhaps more complex than the adoption of human-to-human advice. As algorithms lack transparency, even simple algorithms may be hard for decision makers to interpret and understand (Diakopoulos, 2016), which may diminish users' trust in the algorithm and result in lower rates of adoption. Despite its importance, to date, little research has examined exactly how people adopt or reject AI systems.

Recently, a group of researchers developed the algorithm acceptance model as a conceptual and integrative framework of user-centered algorithms based on the TAM, an information system theory modeling how users come to accept and use a technology (Davis, 1989). Despite being extensively used, the TAM is limited in its ability to examine the complicated processes of user information processing for emergent systems, such as AI and algorithms (Tamilmani et al., 2019; Shin, 2019). To overcome the simplicity of the TAM, several researchers (Shin et al., 2020) have proposed an algorithm acceptance model incorporating antecedent variables of perceived usefulness and convenience or ease of use (Figure 1.1). When users face AI systems and algorithms, several factors affect their decisions

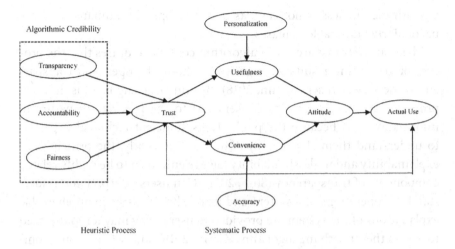

FIGURE 1.1   Heuristic-Systematic Model of the algorithmic process.

on whether they will use them and how (Dwivedi et al., 2019). Such factors are normally different from traditional technologies. Given the black box nature of algorithms, users perceive them as valuable when their design and services are transparent, fair, and accountable (Shin & Park, 2019), and users perceive them as convenient when the services are dependable and trustworthy (Wolker & Powell, 2021). Users receive benefits from AI services, with the biggest focus being convenience and comfort. In the algorithmic model, fairness, transparency, and accountability are proposed as key antecedents of trust, and these relationships are postulated as users' heuristic process of algorithmic conditions. The model includes accuracy and level of personalization as direct influences on convenience and usefulness, respectively. The model has also been extended to include the explanatory factors of personalization and accuracy as factors used for users' systematic processes of algorithm utilities.

1.4.1 FAccT (Fairness, Accountability, and Transparency)

Algorithmic fairness refers to the principle that algorithmic decisions should not produce biased, discriminatory, or unfair results (Lee, 2018). Inherent in a question about fairness is the implication that algorithms do not always function fairly (Elliot, 2022). The fairness of an algorithm can be judged based on its precision (i.e., the ability to produce precise results), accuracy (i.e., the percentage of correct results), and recall (i.e., the ability to find related results). That said, as mentioned in Section 3, achieving algorithmic fairness is not easy, as it is developed by humans, who are particularly vulnerable to biases.

The issue of transparency in algorithm contexts requires that the process of generating results via algorithms should be open and transparent to the viewers/users (Crain, 2018). By and large, algorithms' internal processes are hardly known to users due to the proprietary nature of information and because the public lacks sufficient technical expertise to understand them (Lee & Boynton, 2017). Recently, the principles of explainability and understandability have been shown to be highly related components of transparency (Shin, 2021). Can users interpret and understand the operations of a system and its results? If easily comprehensible explanations of the system are provided to users, they may forgo the need to access the underlying algorithm (Crain, 2018), and when people comprehend how automated journalism works, for instance, they are more likely to consume the content and trust the embedded algorithms and recommended results (Shin et al., 2022c).

The issue of algorithmic accountability is much more complicated than fairness and transparency. Algorithmic accountability implies that firms should be held accountable for the outcomes of their algorithms (Shin et al., 2022c), including their goals, structures, and functions. Ample cases have shown that algorithms produce undesirable effects but are exempted from responsibility, such as Uber's self-driving car skipping a stop sign and Google's facial recognition program labeling a certain race of people as animals. Such mistakes may be avoided by emphasizing algorithmic accountability (Diakopoulos, 2016).

FAccT are becoming important factors in user acceptance of algorithms, as AI is increasingly raising major ethical concerns, such as bias, privacy, discrimination, surveillance, and the role of human judgment. These three issues are thus becoming critical aspects for human judgments about whether to accept and embrace algorithmic technologies.

1.4.2 Trust and Utility

A group of research has shown that FAccT as antecedent variables influences the users' notion of trust, which then affects users' perceived utility. The existing literature on technology acceptance has found that trust influences the actual usage of a system (Glikson & Woolley, 2020). Extensive studies have consistently reported a proven relationship between trust and subsequent behaviors (Alexander et al., 2018). Typically, when people have trust in a system and its services, they are more likely to use it. Applying this relation to AI services suggests that users would use the algorithm if they trusted it (Glikson & Woolley, 2020). Trust in AI is a common theme in the technology acceptance literature, suggesting that it is unclear to what extent algorithmic results can change behavior or to what extent users trust the results. Whether users trust certain systems or services affects the user's assessment and thus influences the user's willingness to share more data with the systems and services (Glikson & Woolley, 2020). Without trust, algorithms cannot realize their potential value because users' data are critical in running predictive and prescriptive analytics for AI. When people trust AI, they consent to share their data with it. Thus, a lack of trust is the key factor that holds back a wider diffusion of AI. Recent studies (e.g., Lee, 2018) have confirmed that trust in AI is a key cause of user attitudes and behaviors regarding algorithmic technologies.

Trust in AI can be multidimensional and complicated. In the context of recommender algorithms, trust is defined as the belief in the reliability and accuracy of the recommended services and the system's capabilities (Lee, 2018). Thus, trust signifies how reliable and credible the system is.

Trust also influences the utilitarian values of technologies, such as their usefulness, ease of use, and convenience (Zheng et al., 2014). According to Gansser and Reich (2021), the user perception of quality is greatly influenced by trust. That is, when users trust AI, they believe that the quality of the algorithms is high enough to provide them with useful and convenient services (Stahl et al., 2021). Therefore, usefulness and convenience are shaped and constructed within users' cognitions based on the trust they have in AI. As such, the utilitarian values of AI are subjective as much as they are objective, and the notion of how useful and convenient a service is depends upon the user's understanding of trust.

1.4.3 Personalization and Accuracy

Algorithmic systems provide personalized services that deliver content and functionality that match a specific user's interests or needs. Many digital platforms and marketing firms rely on the ability to personalize the customer experience. For example, Netflix embeds filtered personalization algorithms to rank its content order and determine what to show customers on their accounts. Based on user data, Netflix provides accurate solutions from the recommender system and increases overall engagement. Rather than having the algorithms suggest prescriptive decisions, Hulu recommends choice decisions to users. As compared to Netflix, Hulu's recommendations depend much less on AI algorithms and more on a human connection.

In 2020, Hulu began a more enhanced recommendation system, which mapped what the user viewed and what time, and a more improved search engine, giving them multiple options to choose from. YouTube applies a complex algorithm to decide the position of videos in its recommendations and lists and to provide the most relevant and personalized videos to its customers. Among hundreds of data points, the YouTube algorithm takes into account user data on average time spent, videos watched, click-through rates, involvement (comments), and feedback submissions.

The capability to provide hyperpersonalization (e.g., relevant content recommendations, tailored product suggestions, and targeted customized responses) influences algorithm performance. When personalized, users are much likelier to accept and engage with the recommended content. Hyperpersonalized content should be precise and accurate as users anticipate personalized suggestions to match their profiles. Relevant research shows that accuracy/personalization are the dual key criteria shaping a user's perceived value of an AI system (Lee et al., 2019).

Personalization in AI enables services to increase user engagement, improve loyalty and performance, and more completely understand their users. Accuracy in AI ensures that the recommended decision or information is in line with users' preferences. Accurate information is particularly critical for medical informatics in healthcare sectors because relying on a mistaken AI decision could cost the lives of patients. According to Shin (2021), people value accuracy in the inferences used to achieve personalization. The accuracy of inferences matters more when personalization is viewed as fair or transparent; however, if users consider the data type used to personalize the recommendations in a given context as biased, the accuracy of the inference does not improve user attitudes.

Personalization and accuracy are key factors in algorithmic service success (Soffer, 2019). Personalizing the AI service is one of the major hurdles that algorithms face due to its complexity and the challenges in understanding and responding to their needs. Legal constraints are also involved. While users enjoy the benefits of personalized and accurate services, they are also becoming increasingly conscious of their personal data. Recent regulations, such as the General Data Protection Regulation (GDPR), have been enforced to give users control over how their data are analyzed and protected. Thus, it is important to ensure a balance between offering personalized and accurate services that users want and respecting their data and privacy rights (Kim & Lee, 2019).

1.5 DYNAMICS OF ALGORITHMIC CULTURE

Experiencing algorithms involves human sensemaking processes, and heuristic systematic processes are inherently human sensemaking. Sensemaking theories addressing cognitive development and information processing theory have confirmed the theories' usefulness in revealing algorithmic sensemaking (Shin, 2021). The cognitive aspect of algorithmic experience involves how users come to understand algorithms and how their ordinary encounters with algorithms form these sensemaking processes (Shin et al., 2022b). The process of making sense of algorithmic results (i.e., how to assess, validate, and contextualize algorithmic outputs) provides essential components of algorithmic experiences. Meaning created through sensemaking is constructed through a reflexive and iterative interaction of data, code, assumptions, heuristics, and algorithmic results (Elliott, 2021). In this light, forming an algorithmic experience can be seen as a sociotechnical process that addresses how users make sense of the world with and through algorithms. It is essential to consider algorithmic experience in the large context of algorithmic culture as such experience is influenced by culture, society, and identity. The argument for algorithmic culture is in line with the metaphor of algorithms as cultural artifacts (Seaver, 2017). Algorithms afford a noble way of understanding the world and a transformation of the grounds of what information means. This affordance is an epistemic change that also influences the context in which information is applied. A compelling argument is that AI and algorithms are cultural artifacts, and underlying algorithmic processes are ontological representations of what people, values, and discourses are typically like. Maceviciute (2021) viewed algorithmic culture as the way in which the logic of AI and algorithms changes how reality is constructed and experienced.

Understanding the dynamic relationship between algorithms and humans as an algorithmic culture gives insights into algorithmic experience (Seaver, 2017). Algorithms learn about the context and culture of users when making decisions, and recommendation algorithms integrate the procedures of human cognition and social experience within a certain culture. Thus, algorithmic culture is closely related to algorithmic experience because algorithmic experience applies algorithmic processes to sort, classify, and hierarchize humans, establish norms, and reinforce existing ideologies based on the practices of thought, behaviors, and meanings that arise in relation to existing algorithmic culture. Striphas (2015) argued that algorithmic culture involves the constant process through which algorithmic processes are used to form human culture. As cultures are systems of judgment and decision making, values, preferences, and interests are all systems of judging the ideas, objects, practices, and performance of algorithms. AI-based chatbots can be considered a cultural artifact because the content chatbots recommend are the result of a socially contextualized practice based on the users' input and preferences, which are reflections of their cultural values and practices (Shin et al., 2022a). Chatbot developers program codes that can learn to stimulate user judgments about information and content to predict and prescribe which users will need what information. At the same time, users give inputs and values to chatbots for more personalized news that they wish to receive. This co-creative and mutually evolving relationship forges a state in which the culture conforms to the user and strengthens their prevailing values rather than confronting them. Algorithms perform what they are trained by people to do, which indicates that algorithms are not neutral; rather, they reflect the values and beliefs held by those who build them, thus reinforcing stereotypes based on those values (Noble, 2018). Many scholars argue that AI should be considered a culture enacted through the interactions and practices of different users and people. Algorithms feed chatbots to produce new recommendations, codes of conduct, habits of generation, and representations of information. Such an algorithmic culture can reinforce existing preferences or personalized information (Hallinan & Striphas, 2016). Because of this cultural embeddedness, there have been many instances in which AI-driven tools and applications reinforce existing human biases and societal imbalances.

The idea of algorithmic experiences has been used to analyze the mode in which algorithms and interfaces are reciprocally molded by user action. Since algorithms structure what we see and how significantly we think of social issues, they should be viewed as an algorithmic self of our intention. The algorithmic self is a result of self-design mediated by algorithms that use the analytics of user data. Humans consciously and unconsciously construct their own algorithms and modify them to whatever they desire without having to assign this decision to a commercial entity with black box operations and functioning (Reviglio & Agosti, 2022). Experiencing algorithms is a process of building an "algorithmic self" or "algorithmized self" (Bhandari & Bimo, 2020). Consider Facebook's "News Feed" algorithm, which determines what users read in their feeds. Initially, the algorithms used generic rules for showing and ranking content in the news feed. The algorithms then collect implicit data from which posts a user visited, liked, shared, and stayed on to understand how users interact with the presented content. The algorithm's parameters are then updated to reflect what the machine learning has learned about – which content contributes to the programmers' goal or user engagement. In the following user interaction, the algorithm will show a different array of news related to the user's interests to improve engagement, and it may even randomly present another subconscious. In the next iteration, the algorithm might display a slightly different set of items to collect more information about users' hidden favorites. A feedback loop based on serial human-AI interaction runs the adaptation and learning process; thus, what the algorithm displays influences the users' behavior, which in turn influences the algorithm's prescriptive predictions, which then decide its next prediction, and so on.

TikTok's "For You" feature and "Following" page offer personalized algorithms that repeatedly confront users with different versions of their own personas. This algorithmized self is derived primarily from a reflexive engagement with previous self-representations rather than with one's social connections, which is a networked self that is established through the reflexive process of flexible associations with social circles (Bhandari & Bimo, 2020). TikTok's algorithmized self is egocentrically driven and is concerned with the performance and management of self-identity.

1.6 IMPLICATIONS: WHAT YOU SOW, SO SHALL YOU REAP

The algorithm acceptance model illustrates the process and components of algorithmic experience when users encounter, adopt, consume, and interact with algorithmic systems. Extending the existing technology acceptance model, the algorithm acceptance model embraces attitude, perceived value, behavioral intention, and trust. Unlike the well-known technology acceptance model, algorithm acceptance inherently bears sticky issues, such as a lack of transparency regarding how algorithmic results are made, a shortage of mutually co-constructed dialogue, and an obscurity of algorithmic accountability. An important implication of the algorithm acceptance model is that humans are not simply accepting algorithms but are co-creating the results with AI. Shin (2021) showed how people actively find and curate news on purpose instead of having the news find them. The algorithmic acceptance model is consistent with the idea of active users and points to the strategic importance of a user-in-the-loop process to ensure that the algorithmic performance meets users' ethical values and standards. The key is to keep algorithms aligned with user expectations and find the intersection between human needs and AI potential.

Before accepting algorithms, users want to check the reliability and validity of algorithmic systems. FAccT serve as cues that users use to evaluate and determine the qualities of AI, and research shows that people using the heuristics of FAccT are more likely to accept algorithmic systems since they are able to understand the system and establish trust in the algorithms (Shin et al., 2022b). When users start to trust the system, and the acceptance processes beings, FAccT influence perceived usefulness and convenience, which then affect attitude and intention. This process is a continuous process of interactions.

The discussion of algorithms as a human experience contributes to the development of user-centered AI. An understanding of how users come to understand algorithm functionality, how their attitudes are changed, how their behavioral intentions are affected, what cognitive views are held, and what outcomes are derived from the cognitive process is critical. For example, user-centered AI discusses users' perception and understanding of fairness and transparency instead of AI-based notions such as "artificial fairness or artificial transparency." One user's experience can be another user's information. This trend can be true for the AI industry, and the discussions here offer valuable insights into new algorithmic developments. For the developers of algorithm-based recommendations or other similar machine learning services, the implications of the acceptance model can help advance the system's performance and the user's experience of the results. Practitioners in the AI industry can develop a deeper understanding of how factors, such as users' desire for an algorithm to make sense, can lead them to under-adopt algorithms. For those involved in marketing or distributing algorithmically generated recommendations, these findings illustrate what users look for when considering whether to accept the recommendations.

1.7 CONCLUDING REMARKS

Algorithmic experience is the outcome of procedural logic, involving a varied set of algorithmic analytics that collect, analyze, and learn from algorithms' own learning experience. Algorithmic experience is not only constructed by users, but in reference to a set of algorithmic performances and other parameters such as data, interface, and datafication that, at the same time, help to structure and learn from the reflexive practice. As Lemos and Pastor (2020, p. 10) state, "within these algorithmic ecosystems, services are either built on the basis of users' behavior data patterns and applied as computational tools for learning about their consumption on modes."

Algorithmic experience is becoming increasingly important for developing human-centered AI. Creating AI from the perspective of what suits human and societal needs is far more important than pushing what is technically feasible. Algorithms can become powerful decision-making tools, freeing people to do the critical tasks they do best and leaving the rest to AI. However, this is only possible if humans develop a meaningful understanding and appreciation for how we adopt and incorporate algorithmic advice, challenge the bias in both humans and algorithms, and build a more algorithmic, inclusive, and aware decision-making paradigm. Designing meaningful algorithmic experience can be useful to make the system graspable and expose its weaknesses and open the system for improve, and that does not need to be achieved by compromising the key functions of algorithmic systems.

This chapter proposed a theoretical underpinning that examines the relations between perceptions of ethical values, trust, and intention while also discussing ideas of co-creation and user-in-the-loop. Our discussions highlight algorithm acceptance, positing the concept of algorithmic experience as part of the analytic lens for human–algorithm interaction. The algorithmic acceptance model suggests a dual route of influence of FAccT on users' intention in algorithm systems: one route through heuristic user cognitive development and the other through systematic development, educed by the accuracy and personalization of the system. From the heuristic-systematic process, we further argue that algorithmic experience is inherently related to the user's perception of transparency, fairness, and other typical factors of user experience, indicating the heuristic roles of transparency and fairness in developing user experience and trust. Algorithmic experience can influence a user's perception of algorithmic systems in the context of algorithm ecology, proposing new insights into the design of human-centered algorithm systems. This new algorithmic experience framework for algorithm systems contributes to an integrated approach for designing human-centered AI systems. Future work can investigate how to apply human-centered algorithm systems by investigating how to design algorithmic services that are not only precise and personally tailored but also fair, transparent, and trustworthy. More research is needed to address the challenges associated with decision makers using or adopting AI effectively.

REFERENCES

Ajzen, I., & Fishbein, M. (1980). *Understanding attitudes and predicting social behavior.* Englewood Cliffs, NJ: Prentice-Hall.

Alexander, V., Blinder, C., & Zak, P. (2018). Why trust an algorithm? Performance, cognition, and neurophysiology. *Computers in Human Behavior, 89,* 279–288. https://doi.org/10.1016/j.chb.2018.07.026

Alvarado, O., & Waern, A. (2018). Towards algorithmic experience: Initial efforts for social media contexts. *CHI '18: Proceedings of the 2018 CHI Conference on Human Factors in Computing Systems,* April 2018. Paper No.: 286. doi:10.1145/3173574.3173860

Beer, D. (2017). The social power of algorithms. *Information, Communication & Society, 20*(1), 1–13. https://doi.org/10.1080/1369118X.2016.1216147

Bhandari, A., & Bimo, S. (2020). TikTok and the algorithmized self: A new model of online interaction. *AoIR 2020: The 21st Annual Conference of the Association of Internet Researchers.* Retrieved from http://spir.aoir.org.

Bonini, T., & Gandini, A. (2019). First week is editorial, second week is algorithmic: Platform gatekeepers and the platformization of music curation. *Social Media + Society.* https://doi.org/10.1177/2056305119880006

Chaiken, S. (1980). Heuristic versus systematic information processing and the use of source message. *Journal of Personality & Social Psychology, 39*(5), 752–766.

Courtois, C., & Timmermans, E. (2018). Cracking the tinder code: An experience sampling approach to the dynamics and impact of platform governing algorithms. *Journal of Computer-Mediated Communication, 23*(1), 1–16. https://doi.org/10.1093/jcmc/zmx001

Crain, M. (2018). The limits of transparency: Data brokers and commodification. *New Media & Society.* doi:10.1177/1461444816657096

Davis, F. D. (1989). Perceived usefulness, perceived ease of use, and user acceptance of information technology. *MIS Quarterly, 133,* 319.

Diakopoulos, N. (2016). Accountability in algorithmic decision making. *Communications of ACM, 59*(2), 58–62. doi:10.1145/2844110

Diakopoulos, N. (2018). Bots and the future of automated accountability. *Columbia Journalism Review, 23.*

Diakopoulos, N., & Koliska, M. (2017). Algorithmic transparency in the news media. *Digital Journalism, 5*(7), 809–828. https://doi.org/10.1080/21670811.2016.1208053

Duan, Y., Edwards, J., & Dwivedi, Y. (2019). Artificial intelligence for decision making in the era of big data–evolution, challenges and research agenda. *International Journal of Information Management, 48,* 63–71. https://doi.org/10.1016/j.ijinfomgt.2019.01.021

Dwivedi, Y. K., Hughes, L., Ismagilova, E., Aarts, G., Coombs, C., Crick, T., Duan, Y., Dwivedi, R., Edwards, J., Eirug, A., Galanos, V., lavarasank, P. V., Janssenl, M., Jonesm, P., Kark, A. K., Kizginb, H., Kronemannm, B., Lal, B., & Williams, M. D. (2019), Artificial intelligence (AI): Multidisciplinary perspectives on emerging challenges, opportunities, and agenda for research, practice and policy. *International Journal of Information Management, 57,* 101994.

Dwivedi, Y. K., Huges, L., Ismagilova, E., & 16 authors. (2021). Artificial intelligence: Multidisciplinary perspectives on emerging challenges, opportunities, and agenda for research, practice and policy. *International Journal of Information Management.* https://doi.org/10.1016/j.ijinfomgt.2019.08.002

Elliott, A. (2021). *Making sense of AI: Our algorithmic world.* New York: Polity.

Ettlinger, N. (2018). Algorithmic affordances for productive resistance. *Big Data & Society, 5*(1), 1–13. doi:10.1177/2053951718771399

Gansser, O., & Reich, C. (2021). A new acceptance model for artificial intelligence with extensions to UTAUT2. *Technology in Society, 65,* 101535. https://doi.org/10.1016/j.techsoc.2021.101535

Glikson, E., & Woolley, A. (2020). Human trust in artificial intelligence: Review of empirical research. *Academy of Management Annals, 14*(2), 627–660.

Hallinan, B., & Striphas, T. (2016). Recommended for you: The Netflix Prize and the production of algorithmic culture. *New Media & Society, 18*(1), 117–137.

Ismagilova, E., Dwivedi, Y., & Slade, E. (2022). Perceived helpfulness of eWOM: Emotions, fairness and rationality. *Journal of Retailing and Consumer Services.* https://doi.org/10.1016/j.jretconser.2019.02.002

Kim, D., & Lee, J. (2019). Designing an algorithm-driven text generation system for personalized and interactive news reading. *International Journal of Human-Computer Interaction, 35*(2), 109–121. https://doi.org/10.1080/10447318.2018.1437864

Knijnenburg, B., Willemsen, M., Gantner, Z., Soncu, H., & Newell, C. (2012). Explaining the user experience of recommender systems. *User Modeling and User-Adapted Interaction, 22,* 441–504. doi:10.1007/s11257-011-9118-4

Lee, M. (2018). Understanding perception of algorithmic decisions: Fairness, trust, and emotion in response to algorithmic management. *Big Data & Society, 5*(1), 1–16. doi:10.1177/2053951718756684

Lee, B., & Boynton, L. (2017). Conceptualizing transparency: Propositions for the integration of situational factors and stakeholders' perspectives. *Public Relations Inquiry, 6*(3), 233–251. doi:10.1177/2046147X17694937

Lee, M., Kusbit, D., Kahng, A., Kim, J., Yuan, X., Chan, A., Noothigattu, R., Lee, S., & Psomas, A. (2019). WeBuildAI: Participatory framework for algorithmic governance. *Proceedings of the ACM on Human-Computer Interaction, 3,* CSCW, 1–35.

Lemos, A., & Pastor, L. (2020). Algorithmic experience: Action and data practice on Instagram platform. *Contracampo Brazilian Journal of Communication, 39*(2), 1–13. https://doi.org/10.22409/contracampo.v39i2.40472

Maceviciute, E. (2021). Review of: In S. Hristova, S. Hong, & J. D. Slack (Eds.), *Algorithmic culture: How big data and artificial intelligence are transforming everyday life.* Lanham, MD: Lexington Books. *Information Research, 26*(2), review no. R716 Retrieved from http://www.informationr.net/ir/reviews/revs716.html.

Noble, S. (2018). *Algorithms of oppression: How search engines reinforce racism.* New York: NYU Press.

Reviglio, U., & Agosti, C. (2022). Thinking outside the black-box: The case for "algorithmic sovereignty" in social media. *Social Media + Society.* doi:10.1177/2056305120915613

Rogers, E. M. (2003). *Diffusion of innovations.* New York: Free Press.

Rossiter, N., & Zehile, S. (2015). The aesthetics of algorithmic experience. In: Randy Martin (Ed.), *The Routledge companion to art and politics.* New York: Routledge. doi:10.4324/9781315736693-26

Rudin, C., & Radin, J. (2019). Why are we using black box models in AI when we don't need to? A lesson from an explainable AI competition. *Harvard Data Science Review, 1*(2). https://doi.org/10.1162/99608f92.5a8a3a3d

Seaver, N. (2017). Algorithms as culture: Some tactics for the ethnography of algorithmic systems. *Big Data & Society, 4*(2), 1–18. doi:10.1177/2053951717738104

Shin, D. (2020). How do users interact with algorithm recommender systems? The interaction of users, algorithms, and performance. *Computers in Human Behavior, 109,* 106344. https://doi.org/10.1016/j.chb.2020.106344

Shin, D. (2021). The perception of humanness in conversational journalism: An algorithmic information-processing perspective. *New Media & Society.* doi:10.1177/1461444821993801

Shin, D. (2022). Explainable AI: How explainability impacts the human interaction with AI. *Social Science Asia, 8*(2), 1–21.

Shin, D., Al-Imamy, S., & Hwang, Y. (2022a). Cross-cultural differences in information processing of chatbot journalism. *Cross Cultural & Strategic Management.* https://doi.org/10.1108/CCSM-06-2020-0125

Shin, D., Lim, J., Ahmad, N., & Ibarahim, M. (2022b). Understanding user sensemaking in fairness and transparency in algorithms: Algorithmic sensemaking in over-the-top platform. *AI & Society.* https://doi.org/10.1007/s00146-022-01525-9

Shin, D., & Park, Y. (2019). Role of fairness, accountability, and transparency in algorithmic affordance. *Computers in Human Behavior, 98,* 277–284. doi:10.1016/j.chb.2019.04.019

Shin, D., Zaid, B., Biocca, F., & Rasul, A. (2022c). In platforms we trust? Unlocking the black-box of news algorithms through interpretable AI. *Journal of Broadcasting and Electronic Media.* https://doi.org/10.1080/08838151.2022.2057984

Shin, D., Zhong, B., & Biocca, F. (2020). Beyond user experience: What constitutes algorithmic experiences? *International Journal of Information Management, 52,* 102061. https://doi.org/10.1016/j.ijinfomgt.2019.102061

Sloan, R. H., & Warner, R. (2018). When is an algorithm transparent? Predictive analytics, privacy, and public policy. *IEEE: Security & Privacy,* May/June. https://doi.org/10.2139/ssrn.3051588

Soffer, O. (2019). Algorithmic personalization and the two-step flow of communication. *Communication Theory.* https://doi.org/10.1093/ct/qtz008

Stahl, B., Andreou, A., Brey, P., Hatzakis, T., Kirichenko, A., Macnish, K., & Wright, D. (2021). Artificial intelligence for human flourishing – Beyond principles for machine learning. *Journal of Business Research, 124,* 374–388.

Striphas, T. (2015). Algorithmic culture. *European Journal of Cultural Studies*, *18*(4), 395–412.

Sundar, S., Jia, H., Bellur, S., Oh, J., & Kim, H. (2022). News informatics: Engaging individuals with data-rich news content through interactivity in source, medium, and message. *CHI Conference on Human Factors in Computing Systems*, 1–17.

Tamilmani, K., Rana, N., Prakasam, N., & Dwivedi, Y. (2019). The battle of brain vs. heart: A literature review and meta-analysis of hedonic motivation use in UTAUT2. *International Journal of Information Management, 46*, 222–235. https://doi.org/10.1016/j.ijinfomgt.2019.01.008

Wilson, M. (2017). Algorithms (and the) everyday. *Information, Communication & Society, 20*(1), 137–150. doi:10.1080/1369118X.2016.1200645

Wolker, A., & Powell, T. (2021). Algorithms in the newsroom? News readers' perceived credibility and selection of automated journalism. *Journalism, 22*(1), 86–103. https://doi.org/10.1177/1464884918757072

Zheng, L., Yang, F., & Li, T. (2014). Modeling and broadening temporal user interest in personalized news recommendation. *Expert Systems with Applications, 47*(7), 3168–3177. doi:10.1016/j.eswa.2013.11.020

Ziewitz, M. (2016). Governing algorithms myth, mess, and methods. *Science, Technology & Human Values, 41*(1), 3–16.

CHAPTER 2

Algorithmic Awareness

K NOWING HOW ALGORITHMS FRAME our online experiences has become a precondition for effective algorithm design. With the drastic surge of platform algorithmification, it is important to understand users' awareness of the increasingly omnipresent algorithms on the online platforms they use because those algorithms can influence users' critical decisions by filtering, mediating, and shaping their interactions. There is an increasing need to enable people to understand, reflect on, and wisely interact with algorithms. Understanding what algorithms are and taking the right control of data and privacy have thus emerged as being important for algorithmic interactions. Algorithmic awareness is needed more than ever since algorithms permeate human lives in their totality and because fake news and misinformation are rampant. How good is your knowledge of algorithms? Can you distinguish genuine from false information? Algorithmic awareness means not only being able to read and understand code but also being aware of the existence and role of algorithms and the underlying repercussions of algorithms. User cognitive processes of algorithmic awareness offer theoretical underpinnings for human-centered algorithm systems and practical guidelines for the design of algorithms. Professionals should have appropriate training to supervise the algorithms with which they work.

2.1 WHY IS USER AWARENESS CRITICAL IN ALGORITHMS?

Do users outside of technologies and regulations need to understand how artificial intelligence (AI) functions and what it can provide us with realistically? Do they need to know what algorithm decisions are being made about them, how their data is shared/used, and how they are surveilled? Should users be able to recognize how algorithms file and personalize information? These questions have become increasingly relevant, especially as AI currently permeates almost all human lives. For example, algorithms recommend the news we read and the goods we purchase, and they determine almost all areas of our lives and work in which machine learning and data analytics are being built to perform more efficiently and productively. With the rapid growth of AI in our lives, algorithms have become deeply infiltrated into our society, serving as gatekeepers for data gathering, ad selection, content recommendation, and decision-making (Shin, 2021). Algorithms are strong, complex, pervasive, often opaque, and inaccessible, and they are becoming increasingly powerful in our everyday lives (Dwivedi et al., 2019). It may be true that platforms like TikTok and Google know their users better than they and their friends do. User algorithms can program their behavior for any outcome. For example, Facebook uses users' Likes to predict sensitive personal data that the users might otherwise keep to themselves, such as political views, personality traits, and sexual orientation (Nishant et al., 2020). Using the deep psychological insights obtained from data on Likes, Facebook's algorithms can precisely predict similar social and psychological traits. Similarly, Naver's news recommendation service is driven by algorithms that prioritize misleading, inflammatory, and socially contentious issues, and Instagram's use of algorithms tends to drive social division. Leaked algorithms from Cambridge Analytica that reportedly influenced the US president's election in 2016 were used as a hidden code to manipulate user behavior on social media (Siles et al., 2020).

Important implications of these algorithms are that machine learning embedded in the platforms learns from people's behavior and the information we share online, such as the data we enter when downloading apps and services and the information we share about ourselves in diverse online functions. AI systems then analyze the data from users' historical behaviors to predict their intentions. However, this method of prediction is often risky, as it tends to reinforce established patterns of values and rules, automizing stereotyped results to numerous questions. Therefore, algorithms inherently echo the frames made by their creators, from the kinds of data gathered to the forms of results presented. For example, as a way to make a profit, Netflix's designers may program the content recommendation algorithm to amplify the exposure of ads by prioritizing content that will hold users on their platform, and YouTube's programmers may frame search algorithms to prioritize results from favored firms and preferential groups. However, when using information as parameters, the current algorithms limit the options available to people and potentially shape their views of the world. This type of algorithmic determinism is hazardous since it impedes the human need for multiplicity and broad perspectives and disregards the diverse perspectives of our culture and society. The decision-making algorithm process impacts people's lives, as people's algorithmic interactions shape their decision-making (Brodsky et al., 2020). However, understanding and having awareness of algorithms is not as easy as looking into the black box to see what is inside because algorithms are essentially an array of complicated coding that are hard for laymen to understand. Given the magnitude of data that AI has on our society, it is important that users have a clear understanding of what an algorithm is, what it is capable of accomplishing, and the political, societal, and ethical implications of these systems (Shin et al., 2022a). It is critical that we have open and ongoing discussions with people about the reality of AI.

Along with the pace of algorithmic technology development, the use of algorithms for automated personalization processes continues to raise ethical and privacy concerns (Shin et al., 2022a). In particular, algorithms work intangibly behind the interface, tracking user activities and personalizing what users see on their platforms, but consumers do not know what they are or how they function (Hargittai et al., 2020). A concerning effect is that algorithms not only automate the generation of content but also frame and control the content and often mislead it with rooted prejudices (Gran et al., 2021). This effect prompts several important questions, such as, "Can algorithms be understood as part of the open system?" and "How can users be engaged with algorithmic personalization?" These questions expedite the need to theorize algorithmic awareness as an essential component of algorithm practices (Shin et al., 2022b). The growing attention to algorithmic awareness is based on the recognition that people deserve to understand how AI works (Kotras, 2021) and that user awareness should be front and center of algorithm design and practice, which is consistent with the General Data Protection Regulation (GDPR) (Shin, 2021). The increasing importance of algorithmic awareness, inter alia, is the explicit channel by which users should be more informed about the inner workings of algorithmic systems (Hamilton et al., 2014) since algorithmic awareness can contribute to users changing their information-sharing behavior (Sohn & Kwon, 2020). Thus, topics related to algorithmic awareness, such as fairness, accountability, transparency, and explainability (FATE), have led to vast public debate and have highlighted the urgency of operationalizing and managing this awareness in AI development (Zarouali et al., 2021). FATE has been used as ethical guidelines as well as strategies and operating principles designed to outline how we can balance the inherent bias coming from algorithms and achieve a sense of fair representation while effectively maintaining a high degree of value. One of the suggested measures to manage these issues is to approach FATE from a user perspective to design an algorithm user-centered system. As the issues are related to the black box processes of algorithms (algorithmic opacity), it is worthwhile examining how users make sense of these issues, how they become aware of them, and what impact the process of FATE has on user behaviors (Shin et al., 2022a).

The opacity of the algorithmic process has led to the need to further investigate how users experience and perceive algorithms. For example, Shin (2021) called for research into how consumers practice algorithms and the extent to which people acquire algorithmic proficiency. An algorithmic awareness design should bear a high level of transparency in algorithmic processes to promote a more informed decision regarding data sharing and adoption. Shin et al. (2022b) proposed that users' understanding of transparency, fairness, and responsibility is relevant to the acceptance of personalizing algorithms. How users perceive algorithmic attributes, frame algorithmic capability, and evaluate recommendation output and personalization quality are related to awareness of algorithms, which plays a significant part in designing and developing sustainable AI (Zarouali et al., 2021).

Siles and Meléndez-Moran (2021) showed that users' awareness of TikTok algorithms structures their affective dimension of TikTok. That is, user awareness helps with the operation of algorithms by shaping user attachment to TikTok. Awareness affords a context for understanding the capabilities of users, algorithms, and their dynamic relationships. TikTok users can sensibly enact diverse personal tools to sustain the emotion and affect related to individualized recommendations on TikTok. Other studies on AI adoption highlight the significant link of FATE to the privacy issues of algorithmic decision-making (e.g., Fast & Jago, 2020; Gutierrez et al., 2019). These studies have shown that the more users know about FATE, the more positive they are about privacy, which then increases trust and, eventually, affects the decision of self-disclosure of their data to algorithms (Shin, 2021). Another study examined users' awareness of their algorithmic Naver's News Recommendation curation and discovered a lack of awareness about the presence of the algorithms and their processes among users (Shin et al., 2022a). The research shows that incorporating a visible hint (explanatory cues) into the opaque algorithmic feed curation process helps users quickly develop rationale about how the algorithm works, leads them to more active engagement with their algorithmically-personalized feed and strengthens overall confidence of control on the algorithms.

However, it remains to be seen how users come to perceive and process FATE, what comprises user awareness, how it is related to privacy, and how trust facilitates information disclosure. These matters are worth examining since they will not only extend knowledge on FATE and privacy but may also elucidate the relationship between the two and further reveal the process of privacy calculus (risk-benefit analytic ways for users' intention to disclose personal data) on sharing personal information with algorithms. The processes of establishing algorithmic awareness and evaluating privacy are vital to managing algorithm adoption decisions (Gran et al., 2021).

The implications of developing algorithmic awareness in AI provide meaningful paths that are both conceptual and practical. Understanding the cognitive process of algorithmic awareness and associated factors contributes to the ongoing development of human–AI interactions (Cotter & Reisdorf, 2020) by clarifying user requirements, usability, and values while reducing AI complexity, opaqueness, and systematic bias. While the effect of algorithmization (often called datafication) on user behaviors has been much debated by academics, the empirical understanding of this effect is still limited, particularly from the user's perspective on specific factors such as privacy and self-disclosure. The key takeaway of our discussion is the conceptual groundwork that is needed for algorithmic awareness to support the user privacy calculus process. Practically, the core role of algorithmic awareness lends an alternative direction to designing human-centered AI to avoid dehumanizing trends in algorithmic practices (Swart, 2021).

2.2 KNOWING ALGORITHMS

As the embeddedness of AI is becoming increasingly sophisticated, it is becoming increasingly difficult for users to demystify byzantine algorithms and make informed choices (Ahmad et al., 2020). When users lack knowledge about AI and the algorithms, they may have an incorrect image of how their and others' data shape their personalized news feeds, for example. Such issues, along with the power of opaque algorithms in shaping users' platform experiences, further raise questions on how informed users are and how informed experience should be incorporated into these algorithms. Numerous misunderstandings have arisen about how these algorithms function and what they can realistically achieve. Algorithmic awareness has become a precondition for algorithmic normative values such as transparency, fairness, and accountability. Thus, getting the right notion of what algorithms are and what they are capable of becoming is a key prerequisite for algorithmic literacy. After many years of the presence of algorithms in our society, people have developed some forms of algorithmic awareness, but people still lack the literacy to understand them.

Awareness matters because it shapes user behavior (Logg et al., 2019) and makes users comprehend algorithm quality (Alter, 2021). In algorithms, awareness breeds particular ways of understanding and interacting with AIs. Arising questions include how users cognize the mechanisms of algorithm platform personalization, how they come to have efficacy through awareness, and what the implications of their efficacy are for assessing privacy and self-disclosure. We should broaden our understanding of how humans make sense of algorithmic output in reference to data, process, and user cognition by taking into account issues of privacy in the algorithmic awareness process.

Regarding discourses involving FATE, algorithmic awareness is more aligned with laymen people's understanding of algorithmic systems. As with FATE, the concerns of the discourse are illustrative of pressing sociocultural, economic, and political needs because a series of concerns related to the functionality of the algorithms are associated with how these functions are sociotechnically and socio-politically deployed within the social world. However, algorithmic awareness continues to be constrained by the lack of a conceptual definition and operational measure (Cotter & Reisdorf, 2020). Although algorithmic logics presently underpin the workings of most algorithm platforms, relevant works report that most users do not know that platforms like Netflix embed algorithms to automate their recommendations (Siles et al., 2020). One survey revealed that less than 19% of social media users were aware of the algorithmic mediating of their information feeds (Smith, 2018). In general, people have little awareness of how their data are collected and used or how such algorithmic personalization comes about, let alone that specific methods are put in place to control algorithm-driven practices. While some users were aware of the basic levels of algorithmicizing processes inside AI, this awareness was not reflected in their experiences with AI. Users may be aware of cognate processes without necessarily engaging with the methodological mechanisms related to the algorithms (Gruber et al., 2021). Users may have an intuitive algorithmic awareness, which might seem advantageous at first, but algorithm awareness and corresponding experience should be lined and extended. Algorithm awareness and algorithm literacy can support a meaningful algorithmic experience. As Klumbyte et al. (2020) noted, algorithmic experience should include the critical capacity of users, which is about algorithmic literacy.

In curating what content is considered personally relevant, algorithms play a crucial part in generating the condition for adoption, usage, and engagement in algorithmic life (Koenig, 2020). The hidden role of these algorithms highlights the need to understand more about the user's level of awareness. The interest in this issue of how users make sense of AI and algorithms has recently grown (Shin et al., 2022a). Sundar et al. (2020) showed the cognitive-heuristic processes through which humans became aware of the actions of algorithms. Users' awareness is related to how much they appropriate algorithmic platforms and is the result of an active engagement with the algorithms. Active users develop a sense of knowing and engagement with algorithms through numerous pathways, including quality evaluation of algorithms, trust judgment, and privacy assessment.

Algorithmic awareness helps people evaluate and interact with algorithmic platforms on the basis that informed judgments lead to better decisions and more effective use of algorithmic resources (Gruber et al., 2021). Algorithmic awareness also helps users assess how platforms, providers, and regulators are using these technologies and, thus, enables them to advocate for responsible technology design and use that avoids problematic biases and helps safeguard privacy (Shin, 2022). According to Koenig (2020), algorithmic awareness also engages meaningful efforts to enable more users to impact data flows and perceive if or when they or others are being marginalized. The influence of these efforts may be limited, depending on the extent of the technical knowledge required. Although researchers have paid extensive attention to algorithmic awareness, it has not yet been well examined (Swart, 2021), and numerous definitions have been suggested for the meaning of algorithmic awareness. For example, Grubber et al. (2021) defined algorithmic awareness as the extent to which users are aware of the existence and function of algorithms in a specific context of consumption, whereas Shin et al. (2022b) referred to algorithmic awareness as the appreciation of how algorithms are used, what they are, how they can benefit people, and how they can negatively impact certain groups. An important hurdle for algorithmic scientists is that AI systems are proprietary and not disclosed to the public. Such limitations make it challenging to identify objective measurements of awareness. While the scientific definition of algorithmic awareness is hard to establish and seems to differ considerably among populations, it is possible to examine how users get a sense of algorithmic awareness – that is, the sensemaking process of algorithmic awareness. As algorithmic processes essentially involve the unfolding of human cognition, behavior, and engagement with the algorithmic logic, users' sensemaking process emerges as an algorithmic culture that has a substantial influence on how platforms and people relate to each other (Shin et al., 2022b).

In AI contexts, user awareness matters because it models algorithms and user actions (Cotter & Reisdorf, 2020). Relevant research has consistently reported that when users are cognizant of underlying algorithm logic, their awareness guides how they behave online (Klawitter & Hargittai, 2018). How users consider algorithms and what they understand about AI shape the way they interact with and/or influence each other. While a few studies have investigated user awareness in different contexts, no standardized scale has yet been developed to measure their awareness. A critical implication from prior works is that awareness is the outcome of a dynamic engagement with algorithms; that is, awareness is not static, defined knowledge but rather a process or practice of evaluating the algorithmic attributes that they use and consume (Zarouali et al., 2021). In this light, algorithmic awareness is tied to how much users know about the ethical and normative values of algorithms (Schwartz & Mahnke, 2021).

Considering that algorithmic awareness includes a perception of what algorithms are, knowing where algorithms are used, understanding the intentions and goals of those owning or deploying the algorithm, and taking control of user data and privacy, concepts such as the elements of FATE can be considered factors of algorithmic awareness. Since algorithmic awareness involves critically recognizing the inherent biases and errors in programming (Hargittai et al., 2020), FATE can be an underlying factor that constitutes algorithmic awareness. From the FATE perspective, algorithmic awareness regards participants' understanding of the way algorithm programs filter and process data, recommend social connections, and reconstruct social realities for them. Increasing algorithmic awareness is a necessary counterpart to increased use of the FATE framework in coding as algorithms become embedded in diverse domains of services (Cotter & Reisdorf, 2020).

2.3 ALGORITHMIC SENSEMAKING

Algorithmic awareness is closely related to how users make sense of algorithms – that is, algorithmic sensemaking. The way that users make sense of algorithmic issues in AI has become an important issue, as what algorithms are trying to do is to mimic human cognitive functioning and maximize the proper meaning interpretation of information in a given context. This meaning interpretation is why algorithmic sensemaking is highlighted. Sensemaking is the process that humans use to construct meaning from raw data (Weick, 1995). Different fields consider aspects of user sensemaking, such as communication, visualization, and data analytics. As media platforms have become algorithm-centric, making sense of them all is becoming increasingly critical. Dervin (2003) argued that "sensemaking reconceptualizes factizing (the making of facts which tap the assumed-to-be-real) as one of the useful verbings humans use to make sense of their worlds" (p. 142). Shin (2022) integrated the sensemaking perspective and information processing theory to examine the process that people use to construct meaning from algorithms. Sensemaking theories address cognitive development, and information processing theory has been confirmed to be useful in revealing algorithmic sensemaking (Shin, 2021) by clarifying how algorithmic information is encoded into human cognition and how human cognition perceives the information.

In his book *Sensemaking: What Makes Human Intelligence Essential in the Age of the Algorithm*, Madsbjerg (2019) argued that human sensitivity and creativity are important in the era of AI. Using case studies of large corporations, Madsbjerg (2019) illustrated the importance of human input and cognition in AI automation. According to his argument, in designing AI, we should follow approaches based not only on programming codes and technical data but also on broad social approaches to reflect user perspectives. The importance of user sensemaking is illustrated by numerous business leaders with humanities degrees, such as Ken Chenault (American Express), Michael Eisner (Disney), and Sam Palmisano (IBM). The AI strategies in these firms are well harmonized with sensemaking processes based on liberal arts perspectives Madsbjerg (2019) argued for a multidisciplinary approach to AI strategies because this approach promotes mental dexterity and conceptual, critical, and creative thinking skills for decision-makers. Recognizing new patterns and developing fresh perspectives are critical in a fast-changing and unpredictable era. The technical accuracy, consistency, and efficiency of algorithms should not be ignored in automation, where humans are inferior to robots, but sensemaking is critical in fields where framing, perspective, and scheme are key to success. For example, effective and efficient sensemaking is critical in AI-based decision-making to ensure that the processes involved in the decision are fair and appropriate. Sensemaking is critical in today's AI era. All arenas of healthcare, public service, policy, and education involve the micro analyses necessary for individuals to confidently use AI tools reasonably in order to solve problems.

Despite the potential value of sensemaking, few studies have utilized it as a theoretical frame to examine the algorithmic phenomenon, possibly because sensemaking does not easily lend itself to practical applications. Incorporating an information-processing perspective into sensemaking can be useful for examining user-cognitive mapping for OTT. Combining these two elements, the question is, then, by what process do users give meaning to fairness and transparency. Although sensemaking is often used as an analytical tool undertaken by experts because users must now make expert-like decisions in complex algorithm contexts, it is appropriate to consider the sensemaking process in this context. Previous research on algorithm adoption and users' meaning constructions (e.g., Sundar et al., 2020) revealed the importance of clarifying the sensemaking mechanisms, needs, and values of the users in their local context when interacting with algorithms.

2.4 ALGORITHMIC DECISION-MAKING

Algorithmic awareness takes two forms: appreciation of algorithms and aversion toward algorithms. The two competing behaviors are contrasted clearly in algorithmic decision-making (ADM). With the proliferation of AI technologies, ADM systems are intricately involved in our lives – not only in our everyday communications and interactions but also in high-stakes decisions about our health, finance, and employment (Rahwan et al., 2019). It is expected that algorithms will outperform human judgment in many specialized tasks. Despite its promise, significant skepticism has risen up about ADM among media, users, and society, and distrust has emerged. Such important decisions that were once handled by human experts are now delegated to ADM, raising concerns about bias, social justice, ethics, and human autonomy (Krafft et al., 2020). Such skepticism is grounded in inevitable comparisons to human decision-making (HDM) and the perception that only humans can be entrusted with complex, context-sensitive, and nuanced decisions.

As ADM systems continue to evolve, they must demonstrate that they are worthy of human trust (Fenneman et al., 2021), which is a complex psychological construct involving certainty, vulnerability, and confidence in the decision-maker's integrity, competence, and benevolence (Choung et al., 2022). Previous findings suggest that trust is a key mediator in the adoption of AI (Ferrario et al., 2020), that trust in an algorithm increases when others' use of the algorithm is disclosed (Alexander et al., 2018), and that algorithm adherence depends on its efficacy and the trust that people have in the algorithm (Fenneman et al., 2021). Notably, automated agents are trusted in leadership roles over human agents because of their higher perceived integrity and transparency compared to humans (Höddinghaus et al., 2021).

Findings on algorithm appreciation (Logg et al., 2019) and machine heuristics (Sundar et al., 2020) suggest that humans trust computers more than other humans for certain types of tasks. Algorithm appreciation is driven by the assumption that algorithms can outperform humans because they are unbiased, objective, accurate, and tireless (Logg et al., 2019). Although such trust in algorithms bodes well for the future of ADM systems, humans must also be encouraged to exercise caution when trusting algorithms. While too little trust in algorithms that perform well can lead to accidents or erode human performance, too much trust in weak algorithms can also lead to adverse consequences. Trust in ADM, then, is tied to the ethical use of AI. Although legislative actions have not caught up with the expectations of advocates of AI ethics and privacy, many technology corporations and organizations have developed principles and values on their own to guide the development of ADM systems and other AI technologies (Hagendorff, 2020). Adherence to these values promotes trust and encourages the acceptance of AI technologies (Choung et al., 2022).

2.5 ALGORITHM AVERSION AND APPRECIATION

While an algorithm can be appreciated for its potential to be fair, impartial, and trustworthy in some contexts, in other contexts, it may perform poorly on the same criteria, resulting in aversion. Algorithmic appreciation refers to the positive perception that machines are safer and more trustworthy than humans (Logg et al., 2019). People showed this tendency of algorithmic appreciation when calculating numeric appraisals about overloaded data and predicting the popularity of certain content. This tendency becomes clear when tasks become more difficult. People exhibit algorithmic appreciation for analytical tasks, in which they try to understand the relations between complex concepts (Bogert et al., 2021).

Despite the wide adoption of algorithms, they often face resistance. Algorithm aversion is defined as a negative evaluation of an algorithm that manifests in adverse behaviors and attitudes toward the algorithm compared to humans (Mahmud et al., 2022). In other words, humans who are algorithm-averse refuse to take advice from an algorithm, even if they would accept the same advice from human agents. Algorithm aversion is related to people's assumptions about AI. People expect algorithms to be perfect and humans to be imperfect. That is, when humans make a mistake, this is perceived as acceptable and tolerable. However, when algorithms make errors, people consider them unacceptable, as they assume AI is perfect; thus, their reactions are negatively represented as algorithm aversion (Renier et al., 2021). Research on algorithm aversion has revealed that users prefer recommendations for entertainment from human agents rather than AI (Logg et al., 2019). Numerous frameworks have been suggested to describe the reasons for algorithm aversion and system features that might help decrease aversion (Burton et al., 2022). In general, when people see a lack of decision control in AI, it tends to escalate algorithm aversion. Users tend to evaluate AI in more stringent ways than they do humans. A lack of decision control, such as the inability to judge the transparency and fairness of AI, leads to aversion. Several algorithmic characteristics, such as explainability, have been shown to affect how users evaluate algorithms (Shin, 2021). Another reason people exhibit resistance to algorithms is a lack of knowledge of how the algorithms produce their recommendations. Low algorithmic awareness is considered to lead to algorithmic aversion.

People are generally doubtful that algorithms can make correct decisions in fuzzy areas, particularly if the task entangles a seemingly human characteristic like empathy or emotion (Newman et al., 2020). Algorithm aversion increases when the task is more qualitative or subjective and lower for tasks that are quantifiable or objective. Recently, studies have proposed a means to avoid algorithm aversion (Shin et al., 2020). One way to overcome algorithm aversion is to allow humans control over algorithmic processes and decisions (human-in-the-loop). Another way is to provide explanations about how algorithms operate. Human interpretable explanations have been proven to significantly mitigate aversion. Such explanations include how the algorithm works, why it generates a particular recommendation, and how confident it is in its generation (Renier et al., 2021).

Algorithm aversion is limited to specific contexts and is particularly prominent after the observation of an algorithmic error. However, with recent advancements in AI, more studies have reported an appreciation for algorithms, which refers to positive perceptions that machines are securer and more trustworthy than humans (Logg et al., 2019). While the majority of the American public exhibits concerns over the transparency and fairness of using ADM to render real-life decisions (Smith, 2018), a more recent survey of Dutch citizens showed that 55% of respondents believe that AI is fairer, compared to 32% who answered that humans were fairer, indicating a shift toward algorithm appreciation (Helberger et al., 2020). Preference for ADM over HDM has been found in various scenarios, including job applications, college admissions, news recommendations (Thurman et al., 2019), and topics ranging from media to public health and justice (Araujo et al., 2020). These findings are part of an increasing body of literature that points to the increased acceptance and appreciation of algorithms among the public. Algorithm appreciation is driven, in part, by perceptions that algorithms are more rational, objective, and unbiased than humans. Such perceptions align with the notion of machine heuristics – the rule that algorithms are more objective than humans and can perform tasks with greater precision (Shin, 2021). Therefore, people's use of positive stereotypes about the infallibility and impartiality of algorithms works as a mental shortcut when making judgments about ADM, which in turn leads to an appreciation of algorithms.

Between algorithmic appreciation and aversion, some users display paradoxical behavior toward algorithms. Users' attitudes and perspectives often contradict their beliefs and behaviors due to a lack of cognitive understanding of algorithms. Like the privacy paradox, while people care seriously about understanding AI and algorithms, their behaviors are often not in line with their concerns. Users might maintain that they value their data while sharing actual data on algorithmic personalization.

Some researchers have recently identified phenomena in which users claim to value their personal data while their actual behaviors are inconsistent (Kokolakis, 2017). The opacity of how algorithms function makes it challenging for users to understand what they do and what they are.

While people know the existence of algorithms, most people lack the level of algorithmic literacy required to fully understand the workings and repercussions of algorithmic systems in their lives. Eslami et al. (2015) indicated that when users understand the effect algorithms have on their news feed, they act accordingly. However, this is not always the case when complex algorithms are considered. Algorithm paradox is derived from the privacy paradox, which is the discrepancy between an individual's intentions to value their privacy and how they actually behave in the online environment. It has often been observed that the relationship between customers' intent to disclose personal data and their actual information-sharing behaviors are widely different and inconsistent. This inconsistency has also been observed between people's attitudes toward information transparency features and the actual practice of personalization. This algorithm paradox can be exemplified by transparency, which can help mitigate concerns of fairness, bias, and trust. At the same time, however, it is obvious that disclosures about algorithms pose their own risks: Explanations can be hacked, releasing additional information may make algorithms more vulnerable to attacks, and disclosures can make AI firms more susceptible to lawsuits or regulatory enforcement. This problem can be described as a transparency paradox, where maintaining the visibility of AI may counterintuitively reduce algorithmic performance by inducing those being observed to conceal their activities through hidden codes and other means.

This kind of algorithmic paradox is also found in media platforms. Some algorithmic systems boast intransparency because their inner mechanisms or performance are a business value. This nature of algorithms as business properties can be a paradox: algorithms refuse to be transparent while analyzing as much privacy about users as possible since data form the source that powers AI. A similar type of paradox arises when users realize how algorithmic models are developed. Machine learning is intended to make predictions about users by mapping their profiles to a broader customer base. On TikTok, the For You feed recommends videos based on what similar users go on to view, where "similar" is based on your searching behaviors and patterns. As users spend more time on this feed system, the For You algorithm will adapt to users' usage patterns, honing in on the user needs and preferences over time instead of comparing your patterns with other similar users. Even the most sophisticated machine learning systems only know specific users in terms of how they compare to their peers.

2.6 ALGORITHMIC AWARENESS AND USER HEURISTICS

User-driven processes essentially involve the social and ethical issues of FATE, which emerge as key prerequisites in the design and development of algorithmic systems (Zarouali et al., 2021). Personalized algorithms are programmed to present accurate, customized content (Shin, 2021). How the personalization processes are performed (Monzer et al., 2020), whether the actual recommendations match the user preferences (Klawitter & Hargittai, 2018), how the results are responsible are tied to the issues of algorithmic awareness (Dignum, 2019), and whether algorithmic personalization lives up to their emancipatory promises (Siles et al., 2020) are related to FATE.

Algorithm providers strive to ensure that the results are accurate and precise. Relevant research shows that transparency and fairness play important roles in algorithms by establishing user trust (Shin, 2021). When transparent and fair mechanisms are ensured, users are likely to consider outputs in a more engaging manner. Visibly transparent algorithmic systems can afford users a sense of assurance, and fair recommendations can lead users to a sense of trust. Along with transparency and fairness, users' notion of accountability is found to have a key effect on user attitudes toward algorithms (Diakopoulos, 2016). The accountability of algorithms refers to the principle that algorithm providers should be held responsible for the consequences of their automated algorithms (Shin, 2020). Algorithmic accountability is the need to justify and explain one's decisions and actions to the users with whom the system interacts. For platforms to be considered responsible, users should be able to assume an evaluative process with positive/negative consequences to follow task outcomes. When users feel a certain level of accountability, they search more conscientiously for relevant information, develop stronger rationalizations for choices, pursue more evidence for decisions, and fulfill tasks themselves more often. This enhanced accountability can be considered a mechanism to lessen the potential negative consequences of automated algorithmic processes (Eslami et al., 2015). High levels of accountability may also lead companies to exert extended efforts to justify decisions. Accordingly, users are likely to use and be gratified when algorithms are held accountable for outcomes, which could decrease unwanted outcomes.

User awareness and literacy of why and how a certain recommendation is curated and how their input affects the decision are significant (Zarouali et al., 2021). Open transparency and clear visibility for relevant feedback improve search performance and user trust in algorithmic systems. Studies

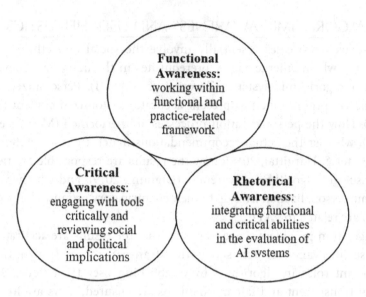

FIGURE 2.1 Algorithmic awareness.

have shown that including explanations can induce positive attitudes and overall satisfaction with a recommendation system (Kizilcec, 2016). Extensive studies have shown a causal relationship between user intention and FAccT in the context of algorithm adoption (e.g., Shin et al., 2022a). Given the ongoing research, it can be inferred that FATE really supports users in understanding the process and thus establishes users' efficacy of the algorithm platforms (Figure 2.1). People like to use explainable systems because they want and have a right to know how data are gathered and processed and, thus, how results are generated (Renijith et al., 2020). When there is a transparent mechanism, users can increase their input data to improve recommendation outputs, and they can understand the logic of a recommendation system (Shin, 2021). The elements of FATE are interwoven and have been found to be associated with the efficacy of systems (Ahmad et al., 2020; Danielsiek et al., 2017).

2.7 YOU CAN SEE AS MUCH AS YOU KNOW

Our discussion contributes to the ongoing development of algorithmic awareness, interaction, and literacy in the context of AI. It is notable to conceptualize algorithmic awareness along with FATE and show the heuristic dimension of algorithmic awareness. Users' self-disclosure levels are positively influenced by the users' awareness level. Also, higher levels of efficacy lead to greater trust and positive privacy perception. These discussions contribute to the theoretical and methodological development of algorithmic privacy by explaining what forms algorithmic awareness take, how algorithmic awareness works, what effects algorithmic awareness has on algorithm use, and from there, how trust can be established, enhanced, and measured. User awareness offers a context for understanding the capacities of humans, AI, and their co-evolving and interdefining relationships (Gutierrez et al., 2019). The contextual factors include the way information is collected by the platforms, how accountable and transparent it is, and the trust perceptions formed through interactions with a specific algorithm. The effect of algorithmic privacy concerns is likely to be outweighed by these contextual factors at a specific level (i.e., related to a specific platform provider in terms of the FATE factors).

Conventional perspectives on awareness have largely remained on the surface of users' knowledge (know-what), leaving their active cognitive process to delve into the roots of an application. As Hargittai et al. (2020) highlighted, it is worthwhile to distinguish algorithmic awareness from technical terms such as coding skills and programming capability (Min, 2019). Unlike technical nitty-gritty, algorithmic awareness goes beyond basic algorithmic skills and includes the hypothetical and contextual awareness of the privacy-trust processes by which algorithms are designed, framed, and consumed, as well as the knowledge that offers users control over these processes (e.g., the right to explain, data control, optimizing privacy, and information). Algorithmic awareness should encompass contextual issues and other inputs to cognize the meaning and implication of data and algorithmic behaviors: users' interpretation of the path algorithms translates meaning, influence and structure our interactions with AIs, and shape the processes of understanding what humans consume, how they derive meaning from algorithms, and what they think. In this light, algorithmic awareness is best framed and practiced as a set of social practices in terms of the ways in which people practice algorithms in their everyday experiences and the interactions that are arbitrated by actual algorithmic services. Algorithmic platforms can be considered experience systems in which their usage and practice enable users to understand how a specific algorithm works (Hargittai et al., 2020). User perceptions and psychological states of mind are critical in rationalizing how and why users perceive and feel what they do about issues surrounding AI, as well as how they accept and experience AI services (Nishant et al., 2020).

The relationship between awareness and trust is a stepping stone to further exploring the role of literacy in the use of AI. We show a more user-centered functionality of how users perceive AI conditions, how their privacy is assessed and processed, what cognitive affordances are realized, and what behavioral results are derived from the processes. Although previous research has consistently shown the role of trust in AI (e.g., Chatterjee et al., 2021), we prove the role of trust in algorithms, their antecedents, their mediating role, and the privacy calculus process. In algorithms, users get a sense of trust when they are assured of algorithmic awareness. When people trust algorithm systems, they tend to believe that the services are safe and useful (Swart, 2021). Trust works as a mediator between user heuristics and quality perception in algorithmic processes. Trust significantly mediates the effects of efficacy on users' self-disclosure behavior. When users trust AI, they tend to disclose more data as they are confident about the data analytic process. Trust between people and algorithms plays a foundational role and permeates the other factors in the process of adoption and experience. The functional features of algorithms are processed through users' sensemaking and processing regarding perceived efficacy, which is mediated by trust. Awareness thus facilitates the cognitive evaluation of norms, performance, attitude, and intention (Eslami et al., 2015). Awareness is having knowledge of privacy, in which you are aware of users and the contextual situation. This point sheds light on "artificial consciousness" as a new area of AI.

2.8 USER AWARENESS BY DESIGN

Our discussions highlight the importance of user awareness in algorithm design and practice. Our algorithmic awareness framework can offer guidelines for informed algorithmic awareness practices that can be integrated into algorithmic platforms. Aspects of FATE have been essential issues in AI, and users seek assurance on such issues when using AI. This study underlines the need to integrate FATE with a design approach when developing algorithm systems and services. A relevant implication of this is that the industry can design innovative "users-in-the-loop" algorithmic platforms to leverage users' capability to deal with algorithmic dilemmas and weaknesses. Developing user-centered algorithm services entails the integration of users' sensemaking processes together with the capability to echo these processes in an algorithm design.

As AI is rapidly developing, the AI industry must develop ways of designing algorithms that are human-aware and user-centric. As AI continues to impact the mode in which we interact with algorithms, how to ensure fair algorithms, transparent interaction, and embed FATE into the interface are pressing issues to resolve. There is a growing demand to inform algorithmic awareness, and those who develop algorithms should be coached in societal ethics and be required to design codes that consider societal processes and their interactions with contexts (Siles et al., 2020). Algorithmic platforms should incorporate a strategy to communicate how their work is fair, transparent, accountable, and in line with accepted social norms to users (Lomborg & Kapsch, 2020). We argue that much effort still needs to be made to ensure an informed public and to narrow the knowledge gap with regard to algorithms and AI. It is necessary to promote awareness of algorithms and AI among all groups in society. We, as users, should also facilitate differentiated media reports aimed at revealing the often unknown applications and hidden effects of algorithms and AI.

2.9 ALGORITHMIC DIVIDE

Algorithmic bias can carry the risk that algorithms and AI may worsen the algorithmic divide – the gap between those able to benefit from algorithms and those who cannot. Different levels of algorithm awareness among users correspond to a new algorithmic divide. Just as the digital divide has divided those with access to digital technologies and digital content from those without, an evolving algorithmic divide currently jeopardizes the reduction of the many social, cultural, economic, political, and various benefits provided by artificial intelligence and algorithms. Conventional discussions of the digital divide have focused on inequalities in the access and diffusion of infrastructure. By clarifying what separates the haves from the have-nots regarding algorithmic awareness, Yu (2020) proposed the concept of the algorithmic divide in a new path by characterizing it according to five elements: appreciation, affordability, access, adaptability, and availability. These elements are in line with different inequalities in knowledge, designing datasets, and treatment, which are tied to the main blocs of social inequalities and, with the rise of AI technologies, affect people's lives and social hierarchy without them even being aware. We conclude that algorithmic awareness is an issue involving humans, algorithms, and interactions, highlighting that algorithms not only enable the flow of information but also structure interactions and determine recommended content. Therefore, we highlight the urgent need to understand whether an awareness lacuna endures among the people in society.

2.10 CONCLUSION

In this chapter, we introduced the concept of algorithmic awareness to illustrate how algorithm users made sense of the functions of algorithmic curations, how they came to form awareness, and the implications of their efficacy through awareness of determining algorithmic trust. An important way for users to acquire a greater influence over algorithmic processes is by engaging in them. Without a fundamental understanding of AI and algorithms, it is impossible for users to use their discretion and critically judge algorithmic procedures and results. Raising awareness of personalization practices is a critical step toward user-centered algorithms. Lacking knowledge of AI and algorithms contributes to exclusion from online discussion, systematic discrimination, diverse privacy risks, and exposure to harmful content. Often underestimated in the current discussion is the active capability of users to make their own choices about what information to search for, which news to follow and reject, and what privacy practices to follow.

Algorithmic awareness goes beyond knowing the technical issues of coding and broadly entails contextually assessing and considering the FATE issues behind the algorithms. Such awareness involves the critical consideration of how algorithms filter content properly, how users can engage with algorithms proactively, how awareness of the intentions and goals contributes to the design of algorithms, and how users are able to control their data and determine individual privacy. Thus, a FATE-aware approach can make algorithmic design more sociotechnically-informed and human-centered.

Future research could probe in greater detail into the methodological and theoretical foundations of FATE and awareness. Making robust theoretical advancements in algorithm awareness is not easy, particularly because this awareness differs widely per user community (Hamilton et al., 2014). One obstacle or future attempts is the reality that neither the researcher nor the user has access to real algorithms to audit and review the influences of different inputs. Despite these challenges, future research can adapt our FATE framework to further investigate algorithmic behaviors in various domains and determine the extent to which user awareness is influenced by/influences algorithmic values and behavior.

REFERENCES

Ahmad, F., Widén, G., & Huvila, I. (2020). The impact of workplace information literacy on organizational innovation. *International Journal of Information Management*, 51, 102041. https://doi.org/10.1016/j.ijinfomgt.2019.102041

Alexander, V., Blinder, C., & Zak, P. J. (2018). Why trust an algorithm? Performance, cognition, and neurophysiology. *Computers in Human Behavior, 89,* 279–288. https://doi.org/10.1016/j.chb.2018.07.026

Alter, S. (2021). Understanding artificial intelligence in the context of usage. *International Journal of Information Management,* 102392. https://doi.org/10.1016/j.ijinfomgt.2021.102392

Araujo, T., Helberger, N., Kruikemeier, S., & de Vreese, C. H. (2020). In AI we trust? Perceptions about automated decision-making by artificial intelligence. *AI & Society, 35*(3), 611–623. https://doi.org/10.1007/s00146-019-00931-w

Bogert, E., Schecter, A., & Watson, R. T. (2021). Humans rely more on algorithms than social influence as a task becomes more difficult. *Scientific Reports, 11,* 1–9. doi:10.1038/s41598-021-87480-9

Brodsky, E., Zomberg, D., Powers, L., & Brooks, J. (2020). Assessing and fostering college students' algorithm awareness across online contexts. *Journal of Media Literacy Education, 12*(3), 43–57. https://doi.org/10.23860/JMLE-2020-12-3-5

Chatterjee, S., Rana, N., Dwivedi, Y., & Baabdullah, A. (2021). Understanding AI adoption in manufacturing and production firms using an integrated TAM-TOE model. *Technological Forecasting and Social Change, 170,* 120880.

Choung, H., David, P., & Ross, A. (2022). Trust in AI and its role in the acceptance of AI technologies. *International Journal of Human-Computer Interaction,* 1–13. https://doi.org/10.1080/10447318.2022.2050543

Cotter, K., & Reisdorf, B. (2020). Algorithmic knowledge gaps. *International Journal of Communication, 14,* 745–765. https://doi.org/1932-8036/20200005

Danielsiek, H., Toma, L., & Vahrenhold, J. (2017). An instrument to assess self-efficacy in introductory algorithms courses. *Proceedings of the ACM Conference on International Computing Education Research,* 257–265. doi:10.1145/3105726.3106171

Dervin, B. (2003). Audience as listener and learner, teacher and confidante: The sense-making approach. In B. Dervin, L. Foreman-Wernet, & E. Launterbach (Eds.), *Sense-making methodology reader: Selected writings of Brenda Dervin* (pp. 215–231). Cresskill, NJ: Hampton Press, Inc.

Diakopoulos, N. (2016). Accountability in algorithmic decision making. *Communications of ACM, 59*(2), 58–62. doi:10.1145/2844110

Dignum, V. (2019). *Responsible artificial intelligence: How to develop and use AI in a responsible way.* New York: Springer International.

Dwivedi, Y. K., et al. (2019). Artificial Intelligence: Multidisciplinary perspectives on emerging challenges, opportunities, and agenda for research, practice and policy. *International Journal of Information Management,* 101994. https://doi.org/10.1016/j.ijinfomgt.2019.08.002

Eslami, M., Rickman, A., Vaccaro, K., Aleyasen, A., Vuong, A., Karahalios, K., & Sandvig, C. (2015). I always assumed that I wasn't really that close to her. *Proceedings of the 33rd Annual ACM Conference on Human Factors in Computing Systems,* 153–162. https://doi.org/10.1145/2702123.2702556

Fast, N., & Jago, A. S. (2020). Privacy matters or does it? Algorithms, rationalization, and the erosion of concern for privacy. *Current Opinion in Psychology, 31,* 44–48.

Fenneman, A., Sickmann, J., Pitz, T., & Sanfey, A. G. (2021). Two distinct and separable processes underlie individual differences in algorithm adherence: Differences in predictions and differences in trust thresholds. *PLoS ONE*, *16*(2), e0247084. https://doi.org/10.1371/journal.pone.0247084

Ferrario, A., Loi, M., & Viganò, E. (2020). In AI we trust incrementally. *Philosophy & Technology*, *33*(3), 523–539. https://doi.org/10.1007/s13347-019-00378-3

Gran, A., Booth, P., & Bucher, T. (2021). To be or not to be algorithm aware. *Information, Communication & Society*. https://doi.org/10.1080/13691 18X.2020.1736124

Gruber, J., Hargittai, E., Karaoglu, G., & Brombach, L. (2021). Algorithm awareness as an important internet skill. *International Journal of Communication*, *15*, 1770–1788.

Gutierrez, A., O'Leary, S., Rana, N., Dwivedi, Y., & Calle, T. (2019). Using privacy calculus theory to explore entrepreneurial directions in mobile location-based advertising. *Computers in Human Behavior*, *95*, 295–306. https://doi. org/10.1016/j.chb.2018.09.015

Hagendorff, T. (2020). The ethics of AI ethics: An evaluation of guidelines. *Minds & Machines, 30*, 99–120. https://doi.org/10.1007/s11023-020-09517-8

Hamilton, K., Karahalios, K., Sandvig, C., & Eslami, M. (2014). A path to understanding the effects of algorithm awareness. *CHI '14 Extended Abstracts on Human Factors in Computing Systems*, 631–642. https://doi. org/10.1145/2559206.2578883

Hargittai, E., Gruber, J., Djukaric, T., Fuchs, J., & Brombach, L. (2020). Black box measures? How to study people's algorithm skills. *Information, Communication & Society, 23*(5), 764–775. doi:10.1080/1369118X.2020.1713846

Helberger, N., Araujo, T., & de Vreese, C. H. (2020). Who is the fairest of them all? Public attitudes and expectations regarding automated decision-making. *Computer Law & Security Review, 39*, 105456. https://doi.org/10.1016/j. clsr.2020.105456

Höddinghaus, M., Sondern, D., & Hertel, G. (2021). The automation of leadership functions: Would people trust decision algorithms? *Computers in Human Behavior, 116*, 106635. https://doi.org/10.1016/j.chb.2020.106635

Kizilcec, R. (2016). How much information? *CHI 2016*, May 7–12, San Jose, CA. https://doi.org/10.1145/2858036.2858402

Klawitter, E., & Hargittai, E. (2018). It's like learning a whole other language. *International Journal of Communication, 12*, 3490–3510.

Klumbyte, G., Lucking, P., & Draude, C. (2020). Reframing algorithmic experience with critical design: The potentials and limits of algorithmic experience as a critical design concept. *Proceedings of the 11th Nordic Conference on Human-Computer Interaction: Shaping Experiences, Shaping Society*, October, 1–12.

Koenig, A. (2020). The algorithms know me and I know them. *Computers and Composition, 58*, 102611. https://doi.org/10.1016/j.compcom.2020.102611.

Kokolakis, S. (2017). Privacy attitudes and privacy behaviour: A review of current research on the privacy paradox phenomenon. *Computers & Security, 64*. doi:10.1016/j.cose.2015.07.002

Kotras, B. (2021). Mass personalization: Predictive marketing algorithms and the reshaping of consumer knowledge. *Big Data & Society, 7*(2). https://doi.org/10.1177/2053951720951581

Krafft, T. D., Zweig, K. A., & König, P. D. (2020). How to regulate algorithmic decision-making. *Regulation & Governance*, rego.12369. https://doi.org/10.1111/rego.12369

Logg, J. M., Minson, J. A., & Moore, D. A. (2019). Algorithm appreciation: People prefer algorithmic to human judgment. *Organizational Behavior and Human Decision Processes, 151*, 90–103. https://doi.org/10.1016/j.obhdp.2018.12.005

Lomborg, S., & Kapsch, P. (2020). Decoding algorithms. *Media, Culture & Society, 42*(5), 745–761.

Madsbjerg, C. (2019). *Sensemaking: What makes human intelligence essential in the age of the algorithm*. New York: Little Brown Book Group Limited.

Min, S. (2019). From algorithmic disengagement to algorithmic activism. *Telematics and Informatics, 43*, 101251. https://doi.org/10.1016/j.tele.2019.101251

Monzer, C., Moeller, J., Helberger, N., & Eskens, S. (2020). User perspectives on the news personalization process. *Digital Journalism, 8*(9), 1142–1162. https://doi.org/10.1080/21670811.2020.1773291

Newman, D. T., Fast, N. J., & Harmon, D. J. (2020). When eliminating bias isn't fair. *Organizational Behavior and Human Decision Processes, 160*, 149–167. https://doi.org/10.1016/j.obhdp.2020.03.008

Nishant, R., Kennedy, M., & Corbett, J. (2020). Artificial intelligence for sustainability. *International Journal of Information Management, 53*, 102104. https://doi.org/10.1016/j.ijinfomgt.2020.102104.

Rahwan, I., et al. (2019). Machine behaviour. *Nature, 568*(7753), 477–486. https://doi.org/10.1038/s41586-019-1138-y

Renier, L. A., Schmid Mast, M., & Bekbergenova, A. (2021). To err is human, not algorithmic–Robust reactions to erring algorithms. *Computers in Human Behavior, 124*, 106879. https://doi.org/10.1016/j.chb.2021.106879

Renijith, S., Sreekumar, A., & Jathavedan, M. (2020). An extensive study on the evolution of context-aware personalized travel recommender systems. *Information Processing & Management, 57*(1), 102078. https://doi.org/10.1016/j.ipm.2019.102078

Schwartz, S., & Mahnke, M. (2021). Facebook use as a communicative relation. *Information, Communication & Society, 24*(7), 1041–1056.

Shin, D. (2020). How do users interact with algorithm recommender systems? The interaction of users, algorithms, and performance. *Computers in Human Behavior, 109*, 106344. https://doi.org/10.1016/j.chb.2020.106344

Shin, D. (2021). The effects of explainability and causability on perception, trust, and acceptance. *International Journal of Human-Computer Studies, 146*. https://doi.org/10.1016/j.ijhcs.2020.102551

Shin, D. (2022). How do people judge the credibility of algorithmic sources? *AI and Society, 37*, 81–96. https://doi.org/10.1007/s00146-021-01158-4

Shin, D., Kee, K., & Shin, E. (2022a). Algorithm awareness: Why user awareness is critical for personal privacy in the adoption of algorithmic platforms?

International Journal of Information Management, 65. 102494. https://doi.org/10.1016/j.ijinfomgt.2022.102494

Shin, D., Zaid, B., Biocca, F., & Rasul, A. (2022b). In platforms we trust? Unlocking the black-box of news algorithms through interpretable AI. *Journal of Broadcasting and Electronic Media*. https://doi.org/10.1080/08838151.2022.2057984

Shin, D., Zhong, B., & Biocca, F. (2020). Beyond user experience: What constitutes algorithmic experiences? *International Journal of Information Management, 52*, 102061. https://doi.org/10.1016/j.ijinfomgt.2019.102061

Siles, I., & Meléndez-Moran, A. (2021). The most aggressive of algorithms: User awareness of and attachment to TikTok's content personalization. *Paper Presented at the Annual Meeting of the International Communication Association*, May 27–31.

Siles, I., Segura-Castillo, A., Solís-Quesada, R., & Sancho, M. (2020). Folk theories of algorithmic recommendations on Spotify. *Big Data & Society, 7*(1), 1–15.

Smith, A. (2018, November 16). *Public Attitudes Toward Computer Algorithms*. Pew Research Center: Internet, Science & Tech.

Sohn, K., & Kwon, O. (2020). Technology acceptance theories and factors influencing artificial Intelligence-based intelligent products. *Telematics and Informatics, 47*, 101324. https://doi.org/10.1016/j.tele.2019.101324

Sundar, S., Kim, J., Beth-Oliver, M., & Molina, M. (2020). Online privacy heuristics that predict information disclosure. *CHI '20*, April 25–30. https://doi.org/10.1145/3313831.3376854

Swart, J. (2021). Experiencing Algorithms: How Young People Understand, Feel About, and Engage With Algorithmic News Selection on Social Media. *Social Media + Society, 7*(2). https://doi.org/10.1177/20563051211008828

Thurman, N., Moeller, J., Helberger, N., & Trilling, D. (2019). My friends, editors, algorithms, and I. *Digital Journalism, 7*(4), 447–469. https://doi.org/10.1080/21670811.2018.1493936

Weick, K. (1995). *Sensemaking in organizations*. 3rd edition. Thousand Oaks, CA: Sage.

Yu, P. K. (2020). The algorithmic divide and equality in the age of artificial intelligence. *Florida Law Review, 72*, 331–389. https://scholarship.law.ufl.edu/flr/vol72/iss2/4

Zarouali, B., Boerman, S. C., & de Vreese, C. H. (2021). Is this recommended by an algorithm? *Telematics and Informatics, 62*, 101607. https://doi.org/10.1016/j.tele.2021.101607

CHAPTER 3

Algorithmic Nudge

ALGORITHMIC NUDGING VIA AI is becoming a popular practice. Nudge principles have been applied to algorithms so that algorithms maneuver the search results through allusive search recommendations and targeted advertisements, steer recommendations, and mix commercials with information in social media feeds. By using algorithms that work invisibly, nudges can be personalized to individuals, and their effectiveness can be traced and attuned as the algorithm improves from user feedback based on a user's behavior. While convenient and useful, these nudges raise a series of ethical concerns about privacy, information disclosure, manipulation, and tweaking. The potential of a single algorithmic nudging to influence thousands of users instantaneously indicates the need to control nudges in AI. The challenge is how to ensure that algorithmic nudges are used in a positive way and whether the nudge could also help to achieve a sustainable way of life. This chapter discusses the principles and dimensions of the nudging effects of AI systems on user behavior and evaluates how people can, in turn, nudge algorithmic systems to have human-centered results.

3.1 DOES ALGORITHMIC NUDGING MAKE BETTER CHOICES?

Nudge theory is a principle in cognitive economics, behavioral science, and cognitive psychology that argues that positive reinforcement and indirect influences guide the behavior and decision-making of individuals or groups (Sunstein & Thaler, 2014). The theory has gained huge popularity with a book written by Nobel-winning economist

DOI: 10.1201/b23083-4

Richard Thaler. Nudging can afford positive reinforcement to individuals and eventually lead to a particular action or decision. Because of its effectiveness, nudging has been widely used in user insights, system development, and public policy areas. Human–computer interaction (HCI) literature has researched numerous principles that suit an online form of nudging well. For instance, Tsavli et al. (2015) showed that enhanced password visualization tools (illustrating password strength levels) can lead to safer passwords. In the information systems literature, numerous relevant studies have examined nudging in a digital context as a subtle form of using design, information, and interaction elements to guide user behavior in online environments without restricting the individual's freedom of choice (Kroll & Stieglitz, 2021).

Nudging, which is termed algorithmic nudge, is defined as the use of algorithm design components to lead user's behavior in algorithmic mediated contexts (Möhlmann, 2021) and is used widely in AI and machine learning services. Also, known as AI nudges, nudging is an indirect way of framing algorithm behavior by nudging human behavior that has been used increasingly and widely (Juneja & Mitra, 2022) to affect purchasing behavior and improve user service. Retailers use AI nudges to engage with their consumers in a highly personalized context that leads them to make better decisions and choices. Shin et al. (2022) conceptualized AI nudges as the use of cognitive stimuli to influence people's behavior predictably without constraining their choices or modifying their incentives. This conceptualization is based on a previous definition of digital nudging by Weinmann et al. (2016), who defined digital nudging as user interfaces in online decision contexts that influence customer actions, although this definition does not indicate the potential of algorithmic technologies. One of the clear differences from conventional nudges is that algorithmic nudges work more secretly behind the algorithm interfaces than general digital nudges. Due to the black box nature of algorithms, algorithmic nudges function stealthily and covertly without being noticed. AI nudges have been used to solidify the primary functions of algorithms, silently shaping what users see, from the posts in social feeds to the suggested content they see to the advertisements covering their online interface. Algorithmic nudges form part of a large AI system to help the larger system work as planned.

Algorithmic nudging uses AI to influence users' behavior by gently coaxing them toward a preferred choice through visual cues, push notifications, and alert alarms. With drastic developments in AI and machine learning, algorithmic nudges have become much more influential than traditional nudges because algorithms can effectively identify target segments in which to nudge users to make behavioral changes. Curating personalized models is a well-honed skill with so much online data about users' behavioral patterns. Algorithmic nudging helps users evaluate and interact with algorithmic systems on the basis that informed judgments lead to wiser decisions (Zarouali et al., 2021). Used correctly, algorithmic nudges help people assess how platforms, firms, and the government use these technologies and, in doing so, enable them to advocate for responsible technology design and use that avoids biases and protects privacy (Akter et al., 2021). Algorithmic nudges can involve meaningful efforts to empower more users to impact data flows and to perceive if or when they or others are being sidelined. The effects of these efforts may be restrained, depending on the extent of the technical knowledge needed (Schobel et al., 2020).

Platform providers use nudges at different dimensions in AI algorithms, such as news recommending services, content purchasing suggestions, and prescriptive decision-making tools. Although the consequences are still in debate and illusive, researchers have examined the relationship between AI and nudges, arguing that algorithmically personalized results can influence users and often lead to unintended consequences and unwanted habits (Burr et al., 2018; Shin et al., 2022). The relationship between AI and nudges found in many studies (Burr et al., 2018) illustrates how personalized tailored algorithms can use persuasion and psychometrics to affect individual and collective behavior in unintended ways. In this regard, Tufekci (2017) conducted longitudinal data to analyze the effects of algorithms on individuals and concluded that algorithmic nudges put users down an ever-darker rabbit hole. Many recent discussions concur with concerns about the ability of algorithmic nudges to predict user taste. The debate is ongoing on how we manage algorithmic nudges to lead to better consequences and whether we should enforce regulations on liability for negative nudges that lead to bad influence. As nudging intermediaries can amplify the severity of public-related harm, it has been voiced that any form of rules and regulation should respond to unethical nudges with varying guidelines for deciding cases of intermediary liability. How to design nudges in AI that guide people toward better decisions ethically and responsibly remains an open question for algorithmic nudges.

3.2 NUDGES AND ALGORITHMIC AFFORDANCE: FROM BLACKBOX AI TO TRANSPARENT AFFORDANCES

Current literature and industry applications on algorithmic nudges tend to focus on framing user behaviors in certain ways, with insufficient attention paid to how people develop cognitive processing or how users respond to nudges (Shin et al., 2022). These questions relate to the principle of affordance, and thus, it is useful to see algorithmic nudges from an affordance perspective. Algorithmic affordances describe the range of explicit and implicit interaction possibilities that enable users to engage with and control the algorithmic system directly and/or indirectly (Shin & Park, 2019). Affordance-based nudges have been proposed as an alternative to algorithmic nudges to mitigate the weaknesses of nudging practices. An affordance-based approach to algorithmic nudging is to follow users' perceptions, information processing, and cognitive development. The concept of affordances can be used as a conceptual framework to understand the relationship between nudges and people, especially with respect to algorithms and AI. A user affordance-based approach can capture users' needs and stakeholders' ideas and generate design options based on those rather than an algorithm-based approach. Affordance-based nudges can make it clear how the device caused a user's action, and affordance is built into the device to show how users can use it without intervention (Shin & Park, 2019). For example, when users see personally recommended product suggestions from the platform, they want to click them to see if they are really personalized to their preferences.

Despite technical sophistication, algorithms rarely offer a practical means for users to interact with them, as they lack the affordances that would enable users to comprehend them or understand how best to use them to complete their tasks. Affordances help the user identify an object's invariants (e.g., functional properties) relative to the user's capabilities, and algorithmic affordances allow users to take action based on their perceptions of features in their environment, such as fairness, transparency, and accountability. In reality, integrating affordances into algorithms is not easy. If the algorithms used by automation are highly complex and different from human cognition or are not perceived by users, the algorithmic results will not be accepted. To take full advantage of algorithmic nudges, it may be necessary to ensure that the algorithm's opportunities for action are made observable and understandable by its users (Baumer, 2017).

The question of how to design algorithmic nudges that are transparent, accountable, and fair relates to how we understand and reflect users' perceived affordances of algorithms. Algorithmic affordance can be activated by trust through users' understanding of fairness, transparency, and accountability. Although users expect algorithms to produce relevant, personalized, and prescriptive recommendations, the underlying assumption is that users presume algorithms to be fair, transparent, and accountable. When these issues are confirmed by users, algorithmic affordance plays a role in the users' consumption of algorithms. From this relation, it can be inferred that algorithmic nudges based on algorithmic affordance have a positive effect on users' cognition and lead them to take steps to recognize, perceive, use, and adopt the suggested nudges.

The principle of affordance is useful for the theoretical conceptualization of algorithmic nudges. Affordance refers to the users' perception of the utility of an object by understanding its features. Users anticipate algorithms that provide personalized, useful, and accurate results, and such desires come with the assessment of fairness, transparency, and accountability. Algorithmic affordance offers a possible cognitive process for perception in humans and the perceptions of transparency, fairness, and accountability (Shin, 2020). When users trust algorithms, they are likely to continue to interact with them by consenting to their data being gathered by algorithms, and the augmented data enables better prescriptive and predictive analytics (Akter et al., 2021). When users confirm algorithmic processes to be fair, transparent, and accountable, they establish trust in AI. Trust and affordance have a circular relationship that once users trust algorithm services or providers, they perceive that the services are easy to use and adopt, and thus they continue to use them. When trust is established, users would like to continue to use as they trust AI, and they are most likely to accept AI with satisfaction. Relevant research has confirmed that transparency, fairness, and accountability build user trust (Burr et al., 2018). The trust feedback loop is a positive feedback loop that reduces users' concerns over transparency and accuracy, and user satisfaction and intention increase significantly (Figure 3.1). Positive feedback is likely positively related to trust, satisfaction, transparency, fairness, and intention. Such feedback implies the significance of examining the complex cognitive mechanisms related to feedback. The positive feedback loop of trust provides heuristic ideas for designing effective algorithmic nudges.

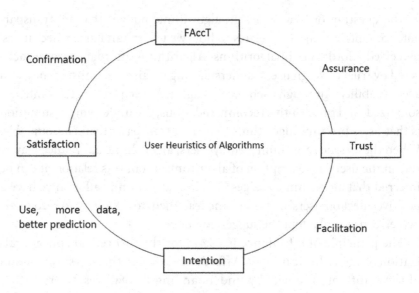

FIGURE 3.1   Positive feedback loop of trust in algorithms.

3.3 ALGORITHMIC SOCIAL MANAGING: ALGORITHMIC BEHAVIOR MODIFICATION

The idea of nudges is not completely new, as the method has been used since the 1980s under names such as behavioral engineering and behavior framing, which were a series of psychological methods to promote better behavior through indirectly encouraging certain behaviors. With AIs becoming mainstream, embedded invisible interventions to encourage people toward desired actions have become more sophisticated. We can easily find numerous examples of algorithms being used to nudge decision-makers toward a particular social outcome. This is often called algorithmic social management. Every day, we face nudges that affect our daily decision-making. Recommender systems are based on algorithmic nudging structures in which content is easily accessible to users and thus shapes their decision-making processes through the automated selection and ranking of presented information. For example, when people buy a burger, they are likelier to purchase fries and drinks if they are offered a bundled recommendation; when there is an additional cost for plastic bags at a superstore, people are less likely to ask for bags, thus decreasing plastic consumption; and colleges hang pictures or posters of leaders and their quotes to encourage students to think in a certain way. Algorithm-based nudges would be even more powerful levers than normal nudges for changing human behavior. AI can be a stable venue for nudging, as people visit the platforms regularly, where they are frequently exposed to numerous choices. As algorithm-based nudges have the ability to learn from data, they can deliver increasingly relevant nudges over time. For instance, Google's Suggested Searches and Autocomplete (autofill) functions are effective kinds of nudging since they gently encourage users to click on the suggested options. Netflix is also developing a kind of nudging by suggesting sequences of episodes at the beginning of menu interfaces. Users could spend minimal effort discontinuing the content, but they were likelier to continue to the subsequent episode due to the kind and quiet suggestion. YouTube also incorporates nudges through the supplementary element of an algorithm-based option advising which content to watch next. Such algorithmic nudges refer to any aspect of the algorithmic architecture that steers and frames people's attitudes and behaviors.

Uber uses algorithmic nudges to maximize the profits and productivity of its drivers (Scheiber, 2021). To create an effective system in which the supply of drivers matches the rising numbers of customers, Uber's algorithms dispatch drivers their next trip ride when the current passenger is still on board. Uber also informs drivers that they are near reaching a particular passenger when the drivers wish to finish their trips for the day, encouraging them to be on the road for longer durations to earn more cash. From the viewpoint of Uber's profit, suggesting the next drive is an effective default setting, as more drivers feel compelled to continue driving. Uber's algorithmic nudge is analogous to the opt-out setting, wherein the drivers have the option of rescinding the notification that pops up by default or continuing with the next ride. The drivers are not required to accept more rides; they are simply being nudged algorithmically by an in-app notification that leads them to the next fare. While drivers can easily opt out with the tap of a button, it is less likely that the driver will select to opt out, as, inherently, people tend to prefer obtaining something to losing out. When the algorithmic nudge leaves very little choice in the hands of the drivers, although Uber is making the decision for the benefit of its drivers, the algorithmic nudge is not always in the best interest of the drivers because they may unintentionally end up working for more rides (Scheiber, 2021). One might argue that those who are contracted to be Uber drivers usually need additional income; thus, Uber's decision to include algorithmic nudges is based on the interests of the drivers. However, it could also be said that Uber is manipulating its drivers by exploiting their needs and economic status. The power dynamic between companies and workers tends to be in favor of the companies. Organizations can use algorithmic nudges to encourage their workers to work more, but ethical issues arise concerning the way in which nudges are utilized and the context in which they are used. It is not the algorithmic nudges themselves that are bad, but the way in which they are implemented could face ethical barriers that reflect on those who use the nudges.

Networked platforms nowadays almost regularly utilize algorithmic behavior modification techniques to shape user behaviors in a way that maximizes profits. Platforms increasingly utilize techniques to flexibly and automatically tailor behavioral interventions to exploit human psychological traits and cognitions. Platforms are created to perform selective gathering, algorithmic filtering, amplification, and monetization of user data and are equipped with autonomous, data-driven, prescriptive, and intelligent algorithms to control users' behavior at scale. People leave their traces online, like digital breadcrumbs through social media posts or pages on Facebook, the Wi-Fi addresses they log into, and the search queries they put in search engines. These digital traces make for a useful and predictive behavioral data reservoir, which platforms use for behavior medication when deciding what news to show, what ads to expose, and what information to avoid. By giving these digital traces to algorithms and platforms, we allow them to infringe on our privacy at times in unexpected ways and fine-tune their recommendation-targeting efforts. Platforms can increase their prediction accuracy by molding users' behaviors toward their predicted values, employing behavior modification techniques, and showing more positive predictions. Such seemingly enhanced predictions can unintentionally result from using reinforcement learning algorithms that integrate prediction and behavior modification.

In journalism, algorithmic nudges are used as a tool to increase news diversity in the news recommender system (Mattis et al., 2022). Loecherbach et al. (2020) illustrated that less-preferred news items are more likely to be chosen if they are displayed at higher ranks or ranked highly on the page. Algorithmic nudges have been proven to considerably increase news diversity, especially when readers experience information overload or when their news selections are recommended by AI systems. Algorithmic nudges could support people who seek but do not find diverse articles by making such articles more accessible and available or by providing unexpected recommendations that prevent unwanted algorithmic feedback loops. These unexpected recommendations lead users to develop new interests, which in turn enables recommendation algorithms to evolve (Karimi et al., 2018), contributing to greater user satisfaction.

Natural language processing algorithms and reinforcing learning are examples of such evolving behavioral changes in algorithms, which are utilized to optimize services and recommendations, promote user engagement, produce more behavioral feedback data, and even hook users via behavioral modifications and long-term addictive habit creation. For example, TikTok's "For You" feed replicates preferences personalized to each user. To hone their personalized For You feed, the service recommends content by rating videos based on a combination of different parameters – starting from the interests users default to as a new user and adjusting for things users specify they are not interested in. In the long run, this service hooks users and forms a habit. In public health, medical, and therapeutic contexts, algorithmic behavior change is a desirable, observable, and replicable intervention intended to modify patient behavior with the participants' consent. However, platform behavioral change methods are becoming increasingly invisible and opaque and are done without clear user approval. Technically, algorithms can program people's behavior for any outcome. A study of the inducive nature of Facebook Likes revealed that Like buttons can precisely forecast personal profiles, such as ethnicity, sexual orientation, political beliefs, religion, and personality. Based on the number of likes, Facebook predicts your friends (150 Likes), your parents (250 Likes), and yourself (300 Likes). It has been criticized as an unethical practice to use such psychological traces acquired from masses of users without their consent to influence users' attitudes. However, the feeds people view on their social media are the results of a specifically directed marketing plan with customized content to nudge a specific user toward a particular action. Although platforms' behavioral change is clear to the user, for example, as presented recommendations, advertising, or auto-complete functions, it is normally invisible and unobservable to third-party auditors. These algorithmic behavior modifications work invisibly and can be operated without external supervision, amplifying filter bubbles, or echo chamber trends.

Former Facebook employee Frances Haugen revealed the significance of transparency and external audits for platforms. In her US Senate testimony in 2021, Haugen stated:

> No one can understand Facebook's destructive choices better than Facebook, because only Facebook gets to look under the hood. A critical starting point for effective regulation is transparency: full access to data for research not directed by Facebook. . . . As long as Facebook is operating in the shadows, hiding its research from public scrutiny, it is unaccountable. . . .

> Left alone, Facebook will continue to make choices that go against the common good, our common good.

This testimony calls for clear platform transparency and access, highlighting the need for a mechanism that conducts independent appraisals and complies with regulatory agencies to audit platform abuse of behavioral modifications, develops appropriate measures to evaluate algorithmic behavior modification impacts, and facilitates public debate in society. Currently, researching algorithmic behavior medication is limited in scope and operation (Greene et al., 2022). The entire interaction of user behavioral data and the associated algorithm data employed for behavior modification and prescriptions are completely unavailable to the public. It is almost impossible to access the operational datasets of platforms that allegedly maneuver behavior modifications. Thus, the role of scholarly researchers and data scientists in this area is unclear. Greene et al. (2022) argued that significant barriers exist between academic researchers and platform operators in terms of legal, scientific, and methodological issues. Therefore, we need to break down the barriers to move beyond academic research practices and develop human-centered algorithms because the collective social costs of algorithmic harm in the era of algorithmic behavior modifications are too huge to overlook.

The European Commission conducted an audit of the Google online shopping recommendation case. The Commission ruled Google in breach of Article 102 because of Google's favorable positioning and distorted display of its own comparison shopping service compared to competing comparison shopping services (Zingales, 2018). Google was fined 2 million euros and directed to take corrective measures to stop the conduct immediately and to avoid engaging in any conduct or act with the same or a similar object or effect. As a result of certain algorithmic design frames, a dominant undertaking smoothly nudges users toward its own services and brands in online shopping, some of which are negative. For example, recommending unhealthy junk foods when shopping online can be a negative nudge since it induces consumers to purchase something harmful to their health and wellbeing.

Realizing the negative effects of algorithmic nudges, in 2021, the Information Commissioner's Office of the UK (ICO) announced a new law prohibiting the use of nudging maneuvers to modify children's behavior negatively. The law stipulates a set of criteria for age-appropriate design that targets safeguarding the personal data of children and prohibits the use of nudge methods to encourage or lead children to offer harmful personal data, deteriorate or close off their privacy safeguards, or extend their use. The Behavioral Insights Team under the ICO has been monitoring algorithmic nudges to continually assess how well organizations are performing.

Algorithmic nudging emerges as a risky means in need of statuary rules and regulations. The behavioral changes via algorithmic nudge can amplify the severity of freedom-related harm. The law should respond to malicious nudges with differential guidelines for deciding cases of intermediary accountability. The question, then, is whether the law should recognize liability for malicious nudges that result in negative consequences. To answer this question, some researchers have proposed "society-in-the-loop" systems (Shin et al., 2022), while others have proposed methods to audit algorithms (Brown et al., 2022). Both approaches seem feasible as long as the general users can control the internal AI code. Alternatively, algorithms should be made that persuade people to behave ethically and conscientiously, although this notion seems to be a long way off.

3.4 CONCERNS OVER ALGORITHM-DRIVEN NUDGES

With its prevalence and wide applicability, there are hot debates and controversies about algorithmic nudging. While effective in monetizing user data, concerns rise as algorithmic nudging has underlined the assumption of manipulative media maneuvers. Algorithmic nudging can be viewed as a new means of the art of propaganda or maneuvered persuasion because it blurs the distinction between free human choice and nudges users toward specific outcomes and behaviors. This negativity is contrary to the argument by Sunstein and Thaler (2014), who initially highlighted the positive roles of nudging in making further decisions about wellbeing, happiness, and health. Despite the generally positive impact of nudges, it has become apparent in the debate that nudges are problematic on many grounds. The problem of influencing people's behavior is that modifying their behavior faces ethical issues and often violates basic human rights. Many critics claim that nudges fail to comply with normative ethical standards and that algorithmic nudges deserve thorough ethical scrutiny (Greene et al., 2022). As the intelligent manipulation of individual behavior on a mass scale is almost a technological reality, algorithmic nudging has been met with as much fierce criticism as it garners great enthusiasm (Shin et al., 2022). While the distinction between manipulation and nudge can be fuzzy, the design of nudges to influence users' decisions in a certain way raises ethical issues. As platforms like Facebook and Instagram continue their push toward algorithmization, we are fast reaching a point where algorithmic nudges will increasingly not only guide us toward the most profitable content for their financial interest but will also remove content that is estimated to be less lucrative. As platforms aggressively seek to maximize the monetization of users, such algorithmic gatekeeping is an effective way to nudge content recommendations through their servers toward more lucrative issues. The nudges coming out of this algorithmic gatekeeping trigger the concern that the nudges infringe on free will or violate basic personal freedom – i.e., user autonomy. In an environment where AI-driven nudges come into play, users may be unaware that the selection they make has been pre-determined by invisible algorithms. Without users' consent, this process can be considered a manipulation of human free will, as it is the algorithm that determines the range of choices. Related to this, several scholars (Shin, 2021) have criticized the idea that algorithmic nudging conflicts with key ethical values, such as liberty and autonomy.

Several civil critics consider that nudges can weaken people's freedom of choice and are not as liberty-preserving as nudge advocates claim (Juneja & Mitra, 2022; Yeung, 2017). This concern relates to volitional autonomy, in that one's behaviors should reflect the individual's desires, interests, preferences, or goals. When influenced by nudges, people may be misled so that their resulting wishes and deeds are no longer really their own. When people are nudged implicitly and covertly, they are no longer the controllers of their selections, and they no longer mirror their self-directed needs. Algorithmic nudges pull our strings and use tricks to get users to do what algorithms need. When people are nudged implicitly and covertly, they are no longer the controllers of their selections, and they no longer mirror their self-directed needs, which supports the criticism that nudges fail to respect rationality, as they often work through opaque or irrational processes. Even if nudges respect our freedom and promote efficiency and convenience, they tap into our irrational or opaque heuristics and biases, which means that they do not treat us as rational human beings and instead dehumanize us. It has been argued that using algorithmic nudging should be done in a highly cautious and discretionary manner to let people make their own decisions (Yeung, 2017). In this light, Shin (2021) is concerned that nudging can remove users from the ability to make diverse judgments and corrodes their remit for their own decisions. Additionally, when the responsibility for making decisions is taken away from people, their sensible assessment and decision-making abilities cannot progress, which in turn weakens their ethical autonomy (Raveendhran & Fast, 2021). This concern relates to deep learning capabilities and emphasizes that becoming a subject of intelligent mass manipulation is not a pleasant thought (Yeung, 2017). Related to this concern is the ongoing debate that exposure to algorithmic nudges predisposes users to make certain decisions or behave toward specific choices. Is this triggered by malicious algorithmic nudges or intentional abuse of the algorithms behind platforms, or is the phenomenon of algorithmic nudges emerging from the experience of searching and interacting with online systems? In other words, is behavior being suggested by algorithmic nudges, or do algorithmic nudges simply echo already existing predispositions for engaging with such content? Shin (2020) argued that algorithmic nudges do not create prejudices, nor do they have any intention of leading users to read fake news and misinformation. The bias we are seeing in AI is the result of a vicious circle in our own content-seeking and use (Shin et al., 2020). Just like a proverb "What you sow so shall you reap," the bias generated from algorithmic nudge are most likely the result of previous user attitudes and actions.

Increasing concerns over algorithmic nudge have led to the Social Media Nudge Act, proposed by Senator Cynthia Lummis (R-Wyoming) and Amy Klobuchar (D-Minnesota). The bill was motivated as a first step toward tackling the algorithmic amplification of destructive nudges. The Act directs the National Academy of Science, Engineering, and Medicine and the National Science Foundation to examine content-neutral ways to increase friction to content sharing online. The bill also requests scholars to identify possible ways of holding up the diffusion of detrimental content and fake news by requesting that users view an article before tweeting or sharing. The Federal Trade Commission is to regulate nudges and order that Twitter or Facebook put nudge rules into action. The background behind the bill is that many politicians believe that social media has garnered undue profits by hooking users and spreading sensational fake news. Debates on this bill argue that, on the one hand, it could address misinformation effectively, but on the other hand, it could threaten free speech by legally allowing censorship mechanisms as a means of controlling debate and suppressing different views. The bill was triggered by Facebook revealer Frances Haugen, and similar bills have since been introduced. Proposed by Tom Malinowski (D-New Jersey) and Anna Eshoo (D-California), the Protecting Americans from Dangerous Algorithms

Act is designed to control algorithmic amplification. Together, these bills can be seen as concerted efforts to increase transparency and improve the user experience by directing algorithmic nudges.

Against the wide use of algorithmic nudges, there is a societal need to think seriously about the governance and control of such algorithms. Some critics are concerned that algorithmic nudging can fall down as a convenient means in the hands of industry or government to exert tricky control over users and citizens and their lives (Brown et al., 2022). Some proponents (Greene et al., 2022) evoke the idea of neoliberalism in which industry and governments increase their control beyond conventional forms of ruling over people and increasingly depend on algorithms to maximize profits. Industry power over users prevails only if industry power is not properly managed by the public. The concern is that algorithms facilitate such domination since they selectively gatekeep public voices and dodge democratic governance behind the shielded algorithms.

Because of the inherent limited transparency and fairness of algorithms, critics have voiced concerns about the manipulative potential of algorithmic nudges (Shin, 2021). Yeung (2017) argued that algorithmic nudges are dangerous because they are manipulative, undermine human dignity, and decrease the human critical thinking process because algorithms influence users invisibly and in a black box instead of transparently. In cases where algorithmic nudges are used in the public sector, primarily by governments, concern is raised about the aims they promote. How can governments determine people's best interests? Governments should not force their goals and interests onto people, particularly not in communities where citizens hold very broad ideas of the good and value.

3.4.1 Algorithmic Un-Nudge: Algorithmic Aversion and Resistance to Algorithms

Within algorithmic nudges, algorithmic behavior is defined as a change in any algorithmic treatment, intervention, or manipulation of platforms designed to influence user attitude and behavior. User behaviors include clicking ads, buying items, posting specific information, and retweeting fake news. Behavior change methods stem from concepts of behavioral psychology and include nudging and operant conditioning. Algorithmic behavior has largely two representations: appreciation and aversion. Algorithm appreciation describes people who are happy with algorithms and are often likelier to accept advice from an algorithm than from a human (Shin et al., 2020). By contrast, algorithm aversion describes a negative assessment of an algorithm that leads to negative attitudes and behaviors toward the algorithm compared to a human service (Logg et al., 2019). Humans often reject recommendations from an algorithm in cases where they adopt the same suggestions coming from human agents (Möhlman & Henfridsson, 2019). However, snubbing advice from algorithms can lead to low performance or a decline in quality (Shin, 2020) because it debilitates the feedback loop due to users and algorithms. It is important to understand why and in what contexts users exhibit algorithm aversion to leverage the benefits of algorithmic nudges to the fullest.

The growing availability of algorithms has increased the trend of algorithm aversion. Researchers have examined the factors that lead to algorithm aversion (Logg et al., 2019). Their findings showed that design factors related to the design of AI lead to aversion. For example, the limited transparency with the black box nature of algorithms means that people cannot understand how algorithms function and produce results. Users are naturally interested in understanding the fundamental logic of the algorithmic process and are thus more inclined to talk to human agents because they can ask questions and understand the rationales behind the results. As such, people turn away from algorithms when they do not understand the algorithmic processes behind the results.

The researchers also found that decision factors can lead to algorithmic aversion (Shin, 2020). For example, factors related to the quality of algorithmic decisions, such as precision, accuracy, and relevance, play a key role in determining user aversion. Inaccurate or irrelevant decisions from algorithms lead people to lose trust in them, as it makes users believe that algorithms are ineffective at completing complicated jobs.

The third cause of algorithmic aversion was found to relate to whether the decisions led people to benefit or suffer from the algorithmic results. The role of algorithmic decisions in relation to human decisions is also a valid factor in algorithmic aversion. People tend to depend more on algorithmic decisions when algorithms enhance human decisions as an assisting role to humans.

Finally, personal factors can contribute to algorithmic aversion. For example, people demonstrate algorithm aversion when the precision of the decisions of both humans and algorithms is indistinguishable and when people evaluate their choices against algorithmic recommendations. Some people are naturally antagonistic to AI, regardless of the algorithmic performance and quality of service, due to their personal disbelief of algorithms, and others reject algorithmic decisions based on their evaluation of the performance of these decisions.

Designers and providers of algorithms should comprehend what factors trigger dislike and what factors lead to appreciation. The factors presented in this chapter are not a comprehensive list, but developers should have a broad perspective of the relevant factors that need to be considered while planning algorithmic nudges. Because users often snub algorithms because of their black box feature design, designers should consider opening the black box while improving algorithmic performance.

3.5 ALGORITHMIC NUDGES WITH MEANINGFUL CONTROL AND ALGORITHMIC AUDIT

Algorithmic nudges are useful in human decision-making because they suggest relevant information while preserving users' freedom of choice. However, algorithmic nudges also present concerns regarding steered bias, data accuracy, and manipulation. As the algorithmically formed reality increasingly shifts from merely nudging people toward predetermined results to removing all other potential results, the algorithms are seemingly taking over. Social platforms are generally silent on issues regarding the fairness, transparency, and accuracy of their nudging. For example, TikTok has also not responded to questions about its nudge accuracy and legitimacy.

This chapter discusses how algorithmic technologies affect humans and how they steer algorithms to enact choice architectures (e.g., placing profitable services at the best interface position while placing less profitable items in harder-to-click spots) and nudges to influence user behavior. The various principles of algorithmic nudges account for recent developments in AI and machine learning. An important proposition in algorithmic nudges is that humans should remain in control of such nudges. Although algorithmic nudges are increasingly pervasive and embedded in many services and objects, they also create unwanted results where moral responsibility for their nudges cannot be suitably attributed to any particular individual or community. The notion of meaningful user control has been proposed for algorithmic nudges to address responsibility gaps and mitigate negative effects by establishing conditions that give people meaningful control over nudges. Algorithmic nudging should enable users to make better decisions by facilitating cognitive processes, extending engagement to construct data, and augmenting users' abilities to utilize insights from the data. For users to control AI meaningfully, they need appropriate knowledge to evaluate AI, known as algorithmic literacy, and they should be able to experience and interact with algorithms to develop algorithmic experience and algorithmic appreciation. The principle of meaningful control of AI is not as easy as it sounds, as it demands numerous conditions as prerequisites. Current discussions of meaningful control over algorithms are either too abstract to be practiced or too narrowly specific, and the technical conditions of transparency or fairness do not consider normal laymen users or the wider societal context. Nevertheless, meaningful control is not completely ideal. The first step toward meaningful control is to enable algorithmic auditing of the AI system, which makes it more interpretable, fair, and controllable. Algorithmic auditing enables us to protect fundamental rights related to privacy and personal data (Brown et al., 2022). Numerous studies have suggested specific methods of algorithmic audit instruments. For example, Brown et al. (2022) proposed three components: a list of the possible interests of users affected by algorithmic nudges, an appraisal metric that explains key ethically salient features of the algorithmic nudges, and a relevancy matrix that connects the assessed metrics to user interests. Shin et al. (2022) proposed algorithmic audit principles (fairness, transparency, accountability, and explainability) for the ethical evaluation of an algorithmic nudge that could be adopted by regulators and policymakers. In any case, it is critical to pay close attention to the complicated societal context within which algorithmic nudges are used and deployed to prevent algorithmic nudges from moving past the limited perspective of traditional nudging as a simple user interface in AI environments. It is important to design algorithmic nudges as human-driven and user-controlled architectures that contribute to overcoming users' emotional, cognitive, and psychological limits when they make decisions and perform actions that contribute to value co-creation.

As algorithms are unlikely to generate the best outcomes by themselves, humans and algorithms should co-create values and enable the design of algorithmic nudge contexts that promote desirable behaviors. The role of interaction stems from a pervasive nudge that shapes contexts and augments capacities for self-understanding, interaction, and user action. This discussion leads to a deeper conceptualization of algorithmic processes. We argue that algorithm creation relies on an interactive process based on users' motivation and knowledge, the orientation of their interactions, and users' needs. While the user's awareness and literacy are critical to value co-creation, new forms of self-understanding and self-development shaped by AI are also important. We conclude that algorithmic nudging allows users to not only select differently but also act differently in practice. By curating the range of users' possible choices, algorithmic nudging contributes to activating users' motivation, thus boosting satisfaction. Algorithmic nudging enacts an algorithm information process that influences users' agencies and practices.

There is a clear distinction between nudging a particular behavior and compelling a specific choice (Burr & Cristianini, 2019). A good algorithmic nudge can encourage a particular choice, but it needs to: (1) be transparent, making the nudge visible and clear instead of hiding other options, costs, or intentions; (2) show available options, enabling users to make the final choice; and (3) be trustworthy, so that users have a good reason to believe algorithmic nudges are warranted, as they are designed to improve user experience.

These suggestions provide the industry with a new design principle for algorithmic nudges and behavioral modifications. Recent advances in machine learning technologies have enabled the establishment of sociotechnical systems that closely interweave users and their social structures with technologies. With the emerging prominence of algorithmic nudges, the question is how to mediate the tension between human choice and algorithmic nudges. This is a key question in algorithmic nudges.

The AI industry can use algorithmic affordance as a guiding principle for programming algorithmic nudges. Users' understanding and perceptions of algorithms and the ways in which users imagine and expect certain algorithmic affordances affect how they approach technologies (Bucher, 2017). As such, the industry could use algorithmic affordance as a key base on which to create feedback loops of machine learning systems such as Facebook, making user beliefs an important component in shaping overall system behavior. Algorithmic affordance not only benefits users by providing them with opportunities to understand how transparent, fair, and accountable algorithmic nudges are, but it could also help industries establish user trust in and satisfaction with their algorithmic nudging.

Algorithm technologies are forging AI ecosystems and enabling new practices; a broader discussion should examine users' decision-making, taking into consideration the broader AI ecosystems. It is essential to analyze how users and algorithms interact and make decisions within co-constructing AI ecosystems. Users' decisions, intertwined with algorithm technologies, determine such co-construction through taking consequent actions that can reform and renew practices and by following such practices and making them institutionalized. Because of the possible risks of manipulation and ethical issues, algorithmic nudges should be designed carefully and used in discretionary ways. Excessive reliance on algorithms for nudges can have unintended results. While giving AI free rein might create effective and sustainable nudges, without the right motivations and rules for the algorithms, algorithmic nudges could lead people to make poor decisions, making them unethical and unsustainable.

REFERENCES

Akter, S., McCarthy, G., Sajib, S., Michael, K., Dwivedi, Y., D'Ambra, J., & Shen, K. (2021). Algorithmic bias in data-driven innovation in the age of AI. *International Journal of Information Management*, 102387. https://doi.org/10.1016/j.ijinfomgt.2021.102387.

Baumer, E. P. (2017). Toward human-centered algorithm design. *Big Data & Society*. doi:10.1177/2053951717718854

Brown, S., Davidovic, J., & Hasan, A. (2022). The algorithm audit: Scoring the algorithms that score us. *Big Data & Society*, 8(1). doi:10.1177/2053951720983865

Burr, C., & Cristianini, N. (2019). Can machines read our minds? *Minds & Machines*, 29, 461–494. https://doi.org/10.1007/s11023-019-09497-4

Burr, C., Cristianini, N., & Ladyman, J. (2018). An analysis of the interaction between intelligent software agents and human users. *Minds and Machines*, 28, 735–774. https://doi.org/10.1007/s11023-018-9479-013

Bucher, T. (2017). The algorithmic imaginary: Exploring the ordinary affects of Facebook algorithms. *Information, Communication & Society*, 20, 30–44.

Greene, T., Martens, D., & Shmueli, G. (2022). Barriers to academic data science research in the new realm of algorithmic behavior modification by digital platforms. *Nature Machine Intelligence*, 4, 323–330. https://doi.org/10.1038/s42256-022-00475-7

Juneja, P., & Mitra, T. (2022). Algorithmic nudge to make better choices: Evaluating effectiveness of XAI frameworks to reveal biases in algorithmic decision making to users. CoRR abs/2202.02479. *CHI 2022 Workshop on Operationalizing Human-Centered Perspectives in Explainable AI.*

Karimi, M., Jannach, D., & Jugovac, M. (2018). News recommender systems: Survey and roads ahead. *Information Processing & Management*, 54(6), 1203–1227.

Kroll, T., & Stieglitz, S. (2021). Digital nudging and privacy: Improving decisions about self-disclosure in social networks. *Behaviour & Information Technology*, 40, 1–19.

Loecherbach, F., Moeller, J., Trilling, D., & van Atteveldt, W. (2020). The unified framework of media diversity: A systematic literature review. *Digital Journalism*, 8(5), 605–642

Logg, J., Minson, J., & Moore, D. (2019). Algorithm appreciation: People prefer algorithmic to human judgment. *Organizational Behavior and Human Decision Processes*, 151, 90–103.

Mattis, N., Masur, P., Möller, J., & van Atteveldt, W. (2022). Nudging towards news diversity: A theoretical framework for facilitating diverse news consumption through recommender design. *New Media & Society*. doi:10.1177/14614448221104413

Möhlman, M., & Henfridsson, O. (2019). What people hate about being managed by algorithms, according to a study of uber drivers. *Harvard Business Review*. Retrieved from www.hbr.org

Möhlmann, M. (2021). Algorithmic nudges don't have to be unethical. *Harvard Business Review*. Retrieved from https://hbr.org/2021/04/algorithmic-nudges-dont-have-to-be-unethical

Raveendhran, R., & Fast, N. J. (2021). Humans judge, algorithms nudge: The psychology of behavior tracking acceptance. *Organizational Behavior and Human Decision Processes*, 164, 11–26. https://doi.org/10.1016/j.obhdp.2021.01.001

Scheiber, N. (2021, April 2). How uber uses psychological tricks to push its drivers' buttons. *New York Times, Technology Section*.

Schobel, S., Barev, T., Janson, A., Hupfeld, F., & Leimeister, J. M. (2020). Understanding user preferences of digital privacy nudges. *Hawaii International Conference on System Sciences*.

Shin, D. (2021). The perception of humanness in conversational journalism: An algorithmic information-processing perspective. *New Media & Society*. doi:10.1177/1461444821993801

Shin, D., Ibrahim, M., & Zaid, B. (2020). Algorithm appreciation: Algorithmic performance, developmental processes, and user interactions. *2020 International Conference on Communications, Computing, Cybersecurity, and Informatics*, November 3–5, The University of Sharjah, Sharjah, UAE. doi:10.1109/CCCI49893.2020.9256470

Shin, D., Lim, J., Ahmad, N., & Ibarahim, M. (2022). Understanding user sensemaking in fairness and transparency in algorithms: Algorithmic sensemaking in over-the-top platform. *AI & Society*. https://doi.org/10.1007/s00146-022-01525-9

Shin, D., & Park, Y. (2019). Role of fairness, accountability, and transparency in algorithmic affordance. *Computers in Human Behavior*, 98, 277–284. doi:10.1016/j.chb.2019.04.019

Sunstein, C. R., & Thaler, R. H. (2014). *Nudge: Improving decisions about health, wealth, and happiness*. New Haven: Yale University Press.

Tsavli, M., Efraimidis, P. S., Katos, V., & Mitrou, L. (2015). Reengineering the user: Privacy concerns about personal data on smartphones. *Information and Computer Security*, 23(4), 394–405. https://doi.org/10.1108/ICS-10-2014-0071

Tufekci, Z. (2017). *Twitter and tear gas*. New Haven, CT: Yale University Press.

Weinmann, M., Schneider, C., & vom Brocke, J. (2016). Digital nudging. *Business & Information Systems Engineering*, 58(6), 433–436. doi:10.2139/ssrn.2708250

Yeung, K. (2017). Hyper nudge: Big data as a mode of regulation by design. *Information, Communication & Society*, 20(1), 118–136.

Zarouali, B., Boerman, S. C., & de Vreese, C. H. (2021). Is this recommended by an algorithm? The development and validation of the algorithmic media content awareness scale. *Telematics and Informatics, 62*, 101607. https://doi.org/10.1016/j.tele.2021.101607

Zingales, N. (2018). Google shopping: Beware of self-favoring in a world of algorithmic nudging. *Competition Policy International-Europe Column.* Retrieved from SSRN: https://ssrn.com/abstract=3707797

CHAPTER 4

Algorithmic Credibility

H OW RELIABLE ARE ALGORITHMS? The use of algorithms is increasing in every sector of AI society, and correspondingly, there is increased concern about their ethical use. A fundamental issue is whether we should trust what we hear about them and what the algorithm recommends. The credibility issue becomes even more important when algorithms are used in critical domains, such as health care and criminal sentencing. It is important to discern the difference between the credibility of claims made *about* algorithms and those made *by* algorithms. A user's sense of belief that algorithms will function in a robust, constructive, and legitimate manner is critical in human–algorithm interaction. This chapter discusses algorithmic credibility by focusing on how reliable algorithms are and by proposing a dual process of algorithmic information processing.

4.1 WHY DOES CREDIBILITY MATTER IN ALGORITHMS?

AI systems have an inherent limitation: they are not perfect, nor should others expect them to be, as they can often err. AI is never perfectly reliable; neither are human beings. Thus, how can we judge the credibility of AI? The issues of algorithmic credibility and trust are becoming increasingly important, particularly in the critical sectors where reliability and accuracy matter (Kolkman, 2022). However, a recent survey reported that almost 70% of online consumers do not trust AI, and about 80% of platform customers do not believe that AI services work for their best interests (Shin et al., 2022). Additionally, more than half of users were uncomfortable sharing personal data with AI systems (Shin et al., 2022). As we rely more on AI for critical decisions, humans are beginning to question the very basis of the performance, and it is imperative that we clarify our view of the nature of AI, the value and meaning of credibility, and the way we trust and interact with algorithms (Alexander et al., 2018). These issues become even more challenging when we do not know how or why certain decisions are made and specific information is recommended. For example, users of AI-driven chatbots seek to find ways to judge the trustworthiness of AI chatbot-curated news. However, making a judgment becomes difficult if the processes used by the underwriting algorithms and the data used for the analytic process are unknown to the users.

Humans tend to accept AI only when they find it credible (Chawla, 2020). Algorithmic credibility provides users with peace of mind, as they can track how and what AI is being used for and offer insights into the why behind algorithm-based decisions. People reject AI if they do not trust AI algorithms, and if the reliability of AI is no higher than around 95%, users consider it inadequate (Alexander et al., 2018). As people expect AI to be almost perfect, they easily reject algorithms after seeing them err, which is termed algorithmic aversion (Hidalgo, 2021). Humans are highly intolerable of algorithmic errors and expect a high accuracy rate for algorithm-based outcomes. Even at a 95% reliable rate of an algorithm-based decision, people would not trust AI because 5% of errors could result in serious consequences. While algorithm-based actions offer better robustness, accuracy, and predictability than human-based actions, people are much less forgiving when it comes to AI errors than human errors, which is why algorithmic credibility matters.

Much of the algorithmic credibility research has focused on expert and professional users of highly sophisticated and advanced AI systems, such as air traffic controllers, nuclear plant operators, and telecommunications traffic controllers. As these algorithms are normally imperceptible and their internal details are not available to the public, they are referred to as black box processes. Most laypeople users are unable to understand the code within the AI, and most users are unaware of how algorithms perform or what role they play in decision-making (Hidalgo, 2021). However, as AI now permeates everyone's lives in everyday applications, it is no longer only professional operators or specialized experts who are influenced by AI and algorithmic systems. AI is now targeted at normal users, as exemplified by explainable AI, transparent AI, responsible AI, and auditable AI, in which trust and credibility serve as prerequisites for successful design. Explainable AI, for example, is designed to give understandable explanations so that users can understand the logic of the algorithms used.

As AI applications are encoded by humans, they can be vulnerable to the exposure of their designers' biases, which may even lead to biased decisions based on flawed input. Common types of biases in AI include sample bias, confirmation bias, prejudice bias, measurement bias, and exclusion bias. The consequences of intentional or unintentional biases in algorithm systems could result in increased consumer concerns, lower user experiences, decreased sales and profits, and potential discrimination. The black box nature of AI could also lead to unfair discrimination, ethics, and threats to user privacy and security. For example, Apple Card was found to be predisposed against gender, as it provided considerably different interest rates and credit lines to males compared to females.

In much of the current debate around algorithmic credibility in AI, issues of fairness, accountability, and transparency (FAccT) are commonly evoked. The issues of FAccT are complicatedly interwoven into AI-driven services, such as chatbots and overall algorithmic phenomena (Shin et al., 2022). FAccT are seen as a way of gaining trust in AI, with a rising concern that the opacity of AIs may decrease the justification for key decisions made by algorithms (Bishop, 2019; Shin, 2021a). FAccT are considered a tool for ensuring non-discrimination, understandability, due process, and responsibility in algorithmic processes (Tsamados et al., 2022; Kitchin, 2017). Concerns about the potentially discriminatory impact of algorithms call for further research into the risks of encoding bias into AI decisions (Obermeyer et al., 2019; Dörr & Hollnbuchner, 2017). Algorithms exist and perform invisibly behind the interface, learning from users and personalizing what users search online, but users do not know what these algorithms are or how they function (Shin, 2020). These issues, including credibility and trust regarding how we assess and accept AIs, remain crucial to algorithm design in the media domain (Kolkman, 2022). If the AI system's priorities are not aligned with fairness, accountability, and transparency, then the AI system could deliver negative outcomes. If users cannot establish trust in AI, its adoption and practice will not produce results for the very reason AI was created. As algorithms bear great benefits as well as risks, all stakeholders, users, organizations, and policymakers should be concerned about how to establish and embed trust in AI to promote adoption and diffusion. This building trust demands deliberate efforts to establish transparency in AI systems and about the data being fed into the algorithms to assure users that their data are being processed properly and ethically to inform and improve performance. As humans, we use our intuitive and cognitive capacities to determine how and whether to trust somebody. To determine whether we can trust, humans examine body posture, facial appearance, and related information or commonly recall our memories or history, which might lead to bias. AI systems are also likely to be biased since they rely on the historical data that is fed to them; thus, any bias in the data is either echoed or intensified in the future forecasts that the model predicts. It is, therefore, critical to examine the intentional and unintentional consequences of algorithms to quickly detect, classify, and mitigate cognate impacts. Some of the algorithmic bias occurs due to their black box nature, a lack of diverse training data, and the models developed by biased designers.

4.2 ALGORITHMIC CREDIBILITY

How do users establish the quality of the recommendations and the credibility of the algorithmic sources?

Relevant research has confirmed that users are less likely to pay attention to information that they do not believe (Kolkman, 2022). Perceived algorithmic credibility has been considered important for the acceptance and diffusion of algorithmic platforms (Wölker & Powell, 2021). Media platform credibility is necessary because users are unable to verify every recommendation generated. As such, users need to rely on platforms to accurately reflect their needs and preferences. Thus, credibility is at the center of media platforms and can be driven by user perceptions of algorithmic performance.

Shin (2022) defined algorithmic credibility as the extent to which users perceive recommendations from algorithms to be trustworthy; it is a significant predictor of algorithmic users' further actions, such as data collection approval or willingness to adopt the opinion of the received recommendation. Similarly, algorithmic trust is described as reliance on or confidence in algorithmic attributes or quality (Chawla, 2020). In other words, algorithmic trust relies on algorithmic quality, while algorithmic credibility impacts AI's ability to be trusted. In simple comparisons, algorithmic trust is about people's impressions of AI, whereas algorithmic credibility comes from users' logic and reasoning. People consider algorithms to be credible when they see that algorithmic outputs are accurate, precise, predictable, and, thus, prescriptive. However, people develop trust in AI based on their impressions and emotional judgments of the overall AI performance and satisfaction thereof. Researchers have used several perspectives to explain how to judge algorithmic credibility (Chawla, 2020). Shin (2021a) showed that algorithmic credibility is significantly related to FAccT. Notably, Shin (2021a) claimed that credibility is user-centered; it is progressively constructed by the user's cognition and information processing rather than given or provided by the medium/message/source of media. Algorithmic credibility is a confirmation of users' assessment of quality, which is based on FAccT. This perspective is aligned with the nature of algorithms, as they are based on user input and engagement (i.e., user data). It is reasonable, then, to link perceived credibility and algorithmic quality in algorithms.

Logg et al. (2019) examined how users recognize algorithmic features, how algorithmic trust is formed, and how users encounter algorithm systems. The findings largely support the link between perceived trust and algorithmic quality. Their research echoes related research in diverse algorithmic contexts. Wölker and Powell (2021) proposed that users perceive FAccT in their experience with news recommender algorithms, and Klawitter and Hargittai (2018) conceptualized users' understanding and literacy with respect to the effect of algorithm-based media. How users realize algorithmic characters, how they experience algorithm services, and how credibility plays a role in such processes are fundamental for defining and designing chatbot services and future AI-driven media (Bishop, 2019; Ford & Hutchinson, 2019). To address such issues, it is necessary to examine which of the users' cognitive processes assess algorithmic credibility based on their interaction with AI (Wölker & Powell, 2021). Algorithmic credibility can be best understood/practiced as a set of social practices: the ways people use algorithms in their everyday lives and the actual events mediated by real-world algorithmic services. The processes of evaluating algorithmic credibility and human understanding in situations of high uncertainty or complexity are vital to promoting algorithm adoption decisions (Mahmud et al., 2022).

Credibility in algorithmic processes is becoming a key part of algorithmic systems and is likely to become an essential parameter for the development and sustainability of algorithmic societies (Kolkman, 2022). Media outlets are starting to think about how their content is perceived and accepted because ratings of credibility play a key role in viewership and adoption. The concept of algorithmic credibility is typically viewed as a multifaceted notion drawing from numerous aspects of coverage, such as belief, confidence, trustworthiness, balance, separation of opinion and fact, precision, and accuracy. Here, we attempt to link credibility issues to shifts in user preferences, find the underlying factors for credibility, and examine the factors driving users toward increased use of algorithmic platforms.

Research on algorithmic credibility is framed in the field of human–AI interaction and cognitive science.

From a human–algorithm interaction perspective, Shin (2022) identified several cognate perceptions when interacting on algorithmic platforms, including fairness, transparency, accountability, trust, relevance, accuracy, and credibility, that can generate credibility based on the algorithms for their news feed, the source that delivers content and advertisements to users. The common assumption of studies examining algorithmic credibility is that it must be sought in real contexts where the algorithms are used by people. This assumption suggests that credibility in algorithms does not emerge on its own but rather that credibility is cultivated and grown through interactions with people. Shin and Park (2019) coined the term algorithmic encounters to highlight that encounters with algorithmic systems can be framed as lived. As such, algorithmic credibility can be best analyzed as it is found in everyday life. Extending this view, a few scholars (Spiegelhalter, 2020) take the position that algorithmic credibility is evolved in practice instead of approaching algorithmic credibility as fixed objects delimited within specific platforms. Therefore, we can infer that algorithmic credibility and trust are relational and multifaced contextually, and we need to approach them from the relational perspective.

Among the multifaceted perceptions of credibility mentioned above, trust is a key dimension of credibility since it comprises the morality of the source and perceived integrity (Ananny & Crawford, 2018). Information is considered reliable if it looks to be transparent, fair, and responsible. Hence, algorithmic credibility is closely related to how users perceive FAccT issues (Sundar, 2020). While trust expresses confidence in the algorithmic attributes, credibility is the reputation of such algorithms, influencing the likelihood of being believed (Borah, 2014). In the chatbot news context, credibility is the degree to which readers consider recommended news trustworthy (Shin, 2022), and news credibility is a key element that constitutes algorithmic media trust. Chatbot service providers should address how their information is accepted since credibility plays a key role in readership patterns.

This multidimensional concept of algorithmic credibility is highly relevant to the algorithmic media domain, where public trust in a vast array of media channels continues to decrease, driven by trends that the news industry is flooded with disinformation, misinformation, and fake news, as well as ambivalence about news from automated unknown sources (Kolkman, 2022). These realities resonate with how algorithm-based media and trust in algorithms come with serious concerns about FAccT issues. With the advent of AI, increased attention has been given to credibility and trust and to ensuring FAccT provides more publicly responsible and socially accountable media from the perspective of users (Thurman et al.,

2019). Pressing questions arise, including how can we believe algorithmic systems, to what extent can we trust in algorithmic processes, and how can we acknowledge the results of algorithmic services (Bedi & Vashisth, 2014). News recommender systems or chatbots produce low value for users when they do not trust the system (Alexander et al., 2018). Credibility can be established in news feed systems by revealing and sharing how the system performs and reaches decisions and what responsibility is borne by the results of recommendations. For users to rely on algorithms, they should be assured of the issues of objectivity, neutrality, confidentiality, and impartiality (Lim & Heide, 2015). Users should be able to request transparency, as well as financial and legal responsibility, in their encounters with algorithms. Users should assume that the results of algorithmic decisions can be explained in a timely fashion to anyone who may be unfavorably affected so that these users have something to say about the decision outcomes. Algorithm designers may also need to justify how individuals' data are being used, as there has been increasing demand for AI to unlock the structure, functions, and processes of the algorithms used to search for, analyze, and make automated decisions. In reality, however, the complicated nature of algorithm systems makes it difficult to unravel where the data are obtained and how they are used in the context of algorithms.

4.3 TRUSTWORTHY AI

As AI becomes pervasive, building trustworthy AI systems is important for its broad acceptance by society. Establishing trust in AI systems is considered critical for their adoption and appropriate use.

Just as trust facilitates relationships between people, it may also mediate the interaction between humans and AI. Trust in AI is needed to ensure the continuous and sustainable interaction between users and algorithms. Trustworthiness in AI is more an attribute of humans than a property of an algorithm system because humans engender trust in AI and algorithms rather than the other way around. Similar to trustworthiness, reliability is also a property of AI and algorithms. As a human concept rather than an algorithmic attribute, trustworthiness is only possible when humans understand the nature, logic, and processes of AI systems. In this light, trust is a human attribute residing in human cognition instead of the technical properties embedded within algorithms (Alexander et al., 2018; Shin, 2022). Shin (2022) demonstrated that trust is a heuristic for AI that influences dependence and conceptualizes trust as a relational attribute between users and AI, but the main driver of trust remains on the human side.

The notion of trust as a human attribute gives important clues about how we can build trustworthy AI systems. An irony in human trust in AI is that the best way to make AI trustworthy is simply to trust AI. This irony is based on the feedback loop between AI and humans that when users trust certain AI services, they tend to believe the services are made through a transparent and fair process. The users' established trust allows the AI system more access to the users' data because when users have trust and are confident in the system, they are willing to share more data with the system. A higher quantity and quality of data improve predictive analytics, which can help the system produce more accurate search results. Users are gratified with highly personalized results, and higher satisfaction means that users will have greater trust in the system and will be more likely to continue to use and adopt it. Thus, a trustworthy algorithm is a set of rules that enhances trust (Alexander et al., 2018).

This feedback loop renders the idea that users should be the focus of the circle so that AI is designed based on a user-in-the-loop principle. This principle embeds trust into the system by design instead of adding trust after the system is developed. The question, then, is how we can conceptualize, measure, and implement abstract qualities like trust, confidence, and trustworthiness. What is needed to establish credibility and trust between people and AI? Several studies have revealed that the most critical factors for establishing credibility and trust are fairness, transparency, explainability (understandability), responsibility, reliability, and consistency (Van de Poel, 2020), similar to the AI trust model proposed by Shin (2020, 2021a). Shin's AI trust model is confirmed as a theoretical framework for how users perceive trust and its effect on their usage behavior in AI contexts (Figure 4.1).

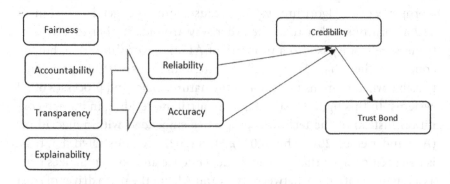

FIGURE 4.1 AI trust model.

This FAccT-oriented credibility model is logical because whenever we encounter algorithms, we must make judgments on whether, how, and to what extent we can trust its algorithmic services. The model has FAccT as the underlying components of trust, which is considered the basis for algorithmic credibility (Swart, 2020). In the model, fairness means that an AI model used for making predictions is not biased with respect to certain attributes such as race, age, and gender; transparency means that the results of an AI model can be suitably explained and discussed so that humans can confirm that the model has been tested and makes sense, and they can understand the rationales of the processes; accountability indicates that humans should be able to identify and assign responsibility for decisions generated by an AI system (Diakopoulos, 2016); and explainability means that an AI model and its output should be explained in a way that makes sense to a human being in an interpretable manner (Rai, 2020). For people to trust AI, they need an explanation for the processed decision or outcome to determine whether the decision is fair, transparent, and accountable, particularly for outcomes or actions that bear negative results.

An important implication is that FAccT serves as an antecedent of algorithmic credibility via accuracy and reliability (Reisdorf & Blank, 2020). The worthiness of belief and trustworthiness of AI services are influenced by how people perceive FAccT in AI systems. When transparent, fair, and accountable services are assured, users are more likely to perceive higher credibility in AI services. High levels of transparent algorithms can give users greater insight into when and why AI algorithms produce personalized results and how to improve their performance. Fair and accountable recommendations afford users a feeling of trustworthiness. For AI to be considered credible, the decisions and outputs must be accurate, consistent, and reliable. Take the example of healthcare facilities that utilize AI to detect defects in ultrasound scans. To be credible and dependable, algorithms need to make accurate and reliable diagnoses, as they may affect someone's life. Accuracy and reliability are the two key measures that define a user's perceived utility of the system. Accuracy refers to whether the personalized system predicts those items that people have already rated or interacted with previously. Algorithmically personalized output is supposed to be accurate, as users presume that the predictive recommendations match their preferences. When users feel that the recommended outputs are optimized to their preferences, they consider the service useful and have confidence in the algorithm (Kim & Lee, 2019). Shin (2021b) confirmed these linkages in a range of algorithm services in which accuracy and personalization were confirmed to lead to credibility and trust.

Output accuracy and technical reliability depend on users' subjective understanding and evaluation of FAccT while interacting with AI systems. Reliability and accuracy characterize whether the predictive results of algorithms are correct and consistent; however, such attributes cannot be measured wholly within the logic of technologies, as they are based on the subjective experience of the users, which reflects the intensity of their emotional, cognitive, and sensory connection to both the content and the modality of algorithms. When users are assured of such reliability and accuracy, they trust the algorithm systems. Established trust enables users to share more data with AI. User understanding of the algorithmic processes is significant in the construction of an algorithm user interface. Numerous studies have shown that including explanations enhances users' trust in and satisfaction with a machine learning system (Shin, 2020). Algorithmic credibility is built through activities in which people understand how algorithms are FAccT. Relevant research has validated the relationship between FAccT and trust (Reisdorf & Blank, 2020). When users have higher algorithmic literacy through FAccT, they will attribute more credibility to chatbot services (Lokot & Diakopoulos, 2015).

The AI trust model incorporates all necessary components of trustworthiness. The model is widely applicable to human–AI interaction, including AI chatbot services, algorithmic recommendation systems, automated decision-making, and information searches. The model is well-aligned with the EU guidelines for trustworthy AI. The guideline proposes three standards: (1) AI should be ethical, guaranteeing adherence to ethical values and principles; (2) AI should be legitimate, complying with all pertinent laws and regulations; and (3) AI should be reliable from both social and technical perspectives. The AI trust model offers a conceptual and strategic way to think about AI credibility and ethics, helping practitioners design, develop, deploy, and run AI systems they can trust.

4.4 AI-BASED CHATBOT INTERACTION: HOW DO USERS INTERACT WITH CHATBOT?

The rise of chatbots in society has opened noble ways of transforming journalistic practices by using interactive dialogue via conversational agents (Araujo, 2018). As an AI-driven conversational agent, a chatbot is an algorithmic program designed to simulate human conversations (Shin, 2021a). Chatbots can help reporters deliver their stories differently or gather information from readers. These systems then produce profiles of user preferences based on prior online behavior and default information (Oh et al., 2021). The use of conversational agents in journalism opens a new window of opportunity, such as conversational journalism (Thurman et al., 2019). The conversational features of the interaction demand that journalistic chatbots present social behaviors. Journalists and news organizations already use chat AI and its ability to make them more conversationally interesting as a way of reaching their readers (Rietz et al., 2019). Chatbots are most commonly used by interactive news services (Jones & Jones, 2019). For example, *Quartz* is a texting service by which users can receive news via a pre-programmed course of messages, and the *New York Times* uses a Slack-bot to suggest news articles about interests specific to users.

As AI services develop, understanding how people interact with journalistic chatbots becomes significant to algorithm design and evolution (Bolin & Schwarz, 2015). In terms of algorithmic features, user heuristics raise the following questions: How do users figure out the media characteristics or features of a chatbot, and how do people perceive and make sense of chatbots? Because algorithm-based content brings a competitive edge and numerous innovative smart services, it is essential to investigate users' a priori expectations and how those expectations are realized. It is also important to understand how users' trust in the algorithm affects their emotions and influences their behavior. Algorithmic information processing can be used as a frame to discover users' cognitive processes regarding chatbot services, as it is proven effective in examining user behavior as a course of sensemaking, experience, and behavior. With chatbot services, we use an algorithmic information processing model to understand the role that algorithmic features play in shaping users' understanding of AI, as well as how users' behaviors affect that understanding.

4.5 ALGORITHMIC INFORMATION PROCESSING: COGNITIVE PERSPECTIVE

Traditionally, algorithmic information theory has been a division of mathematical information science that concerns itself with the relationship between algorithms and the computers of computably generated objects (Burgin, 1990). The theory is mainly focused on the technical perspective of how algorithms process information for automatization. In terms of the human–AI interaction perspective, Shin (2021a) proposed that users' viewpoint of algorithmic information processing focuses on their cognitive development of algorithmic interaction. Several algorithm researchers (Burrell, 2016) describe cognitive processing in terms of the cognitive, psychological, and behavioral changes in a user's mind. Algorithm information processing is designed to understand human cognition in relation to how humans process information from algorithms. This perspective considers human cognition to be essentially algorithmic in nature, with the mind being the code and the brain being the program. This process is based on the idea that users process the algorithmic information they experience instead of merely receiving the algorithmic results. That is, users do not simply respond to algorithms; instead, they actively interact with algorithms to construct the results that they want. This perspective matches the user's view of algorithms based on the assumption that an algorithm is a replication of its human creator in terms of what they feed the algorithm, what they search for, and who they are. Thus, this view analyzes how users perceive algorithmic features and how they process and respond to the algorithmic outputs they receive through their feedback and interactions. Shin (2021a) clarified a continuous and dynamic pattern of development throughout the interaction with algorithms, which is suited to the opaque nature of AI.

The Algorithmic Information Processing theory traces how individuals process the algorithmic information they receive beyond merely responding to stimuli (Figure 4.2). Shin et al. (2022) explained how chatbot users

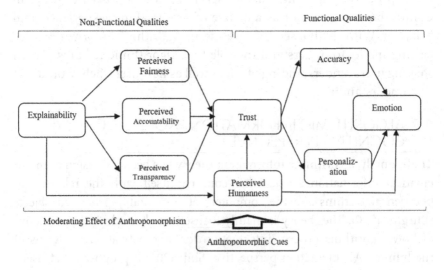

FIGURE 4.2   An algorithmic information processing model.

process different stimuli and how these processes influence their motivations and behaviors. The study clarifies user sensemaking in chatbot news services by elucidating the role that algorithmic features play in framing user perceptions of humanness and the sensemaking of algorithms, as well as how individual actions guide their sensemaking. In his serial works on algorithms, Shin (2022) conceptualized algorithmic factors and proposed two stages of algorithmic information processing: (1) through FAccT heuristics and (2) through the acceptance of the systematic process that proceeds through trust. During this process, users evaluate the features of an algorithm and determine whether to continue using AI services based on these evaluations.

Heuristic processing engages the use of simplifying FAccT decision estimation to quickly make a rough assessment of algorithmic service quality, which individuals must do whenever they encounter algorithms. Systematic processing involves deliberative processing of the practicality and benefits of algorithmic services. Trust connects the two processes, linking the heuristic and systematic mechanisms and providing a key clue to the algorithmic qualities, algorithm experiences, and users' interactions with AI. Certain algorithmic features afford users cues for trust, which allows them to view algorithms with feelings of usefulness and efficacy. It can be inferred that trust shaped through heuristic processing is more likely to have cognitive attributes that reflect the FAccT assessment, whereas trust shaped through systematic processing is more likely to affect performance evaluation due to reliance on established FAccT cues. Credibility plays a facilitating liaison role in the experience of algorithms (Shin, 2020).

Relevant research shows that anthropomorphizing or humanizing AI automation or algorithm-based services can elicit or facilitate algorithmic information processes (Shin, 2021a; Rosenfeld & Richardson, 2019). The findings of these studies reveal a key role of anthropomorphic cues in users' information processes in AI. Another study has confirmed that humanizing anthropomorphic explanations act as heuristic cues, facilitating privacy evaluation and triggering user trust (Araujo, 2018). While the question still remains as to how far we can humanize AI given a possible uncanny valley effect, the facilitating roles of anthropomorphic cues provide significant directions for algorithmic information processing.

As AI services develop, understanding how people interact with chatbot media becomes important to algorithm design and evolution (Bolin & Schwarz, 2015). A comparative case study of AI-based chatbots between the US and Korea shows that a difference exists in the pattern of algorithmic information processing (Shin, 2020). The study showed that the two countries use the same two information processes: heuristic and systematic processing. Individuals who engage in heuristic processing rely on available cues in a persuasion environment and accessible, relevant cognitions in the individual's memory. Individuals who engage in systematic processing obtain a thorough understanding of any available information through comprehensive reviewing, intensive reasoning, and analytic thinking (Bohner et al., 1995).

Users in both cultures who considered AI services to be of usable and high quality largely had positive confirmation, which rendered high emotion, while those who perceived AI to be procedural were more likely to show a positive attitude and approve of transparent algorithmic processes. The findings disclose that users in both cultures differ in terms of how they make sense of their expectations and how they interact with chatbots. Korean users are more oriented toward functional qualities that affect the performance of AI than toward procedural qualities, whereas users in the US are more oriented toward procedural qualities. In other words, for US users, ethical features can work as a heuristic when making evaluations based on the credibility and trustworthiness of AI, whereas Korean users evaluate algorithmic quality in terms of functional features. This difference implies that procedural and performance quality differs by culture, signifying that the needs and values of AI users might also differ across cultures. This result is consistent with previous findings in cross-cultural research, which reported that people in Western cultures tend to use contextualized and analytical approaches to process information compared to users in Asian cultures, who are likelier to use functional evaluations or tangible means when processing information (Shin, 2021a). As a result, US users' preference for normative value might trigger a procedural-motivated evaluation, whereas Japanese participants' preference might result in a performance-favored evaluation. Different attitudes were found to be significant for US users compared to Korean users, and those attitudinal differences relate to a different level of trust and emotional valence. The differences in the patterns also relate to perceived quality. Korean users who perceive the outcome values of AI believe that precision, personalization, and utility are the key factors influencing their emotions. Conversely, US users consider procedural values the key trigger that influences their trust in algorithms. Korean users are more likely to be gratified with the utility of the algorithms than US users, whereas US users are more likely to be satisfied through a heuristic evaluation of procedural qualities than

Korean users. Other underlying factors might explain the use of confirmation by Korean users, and other variables might account for the emotions among US users. Possibly US users expect more than performance in terms of algorithm emotion, whereas Korean users expect more than procedural values in terms of confirmation of procedural qualities. From the findings, it can be inferred that perceived procedural and performance qualities are symbiotic, co-influencing algorithm users across countries.

Further, while US users typically have generalized distrust about the fair process of algorithms, Korean users might care less about the internal process of how the news is curated than the results of the algorithms presented. While US users might not understand the technical algorithm process either, they are aware of widespread societal concerns about ethical issues and skepticism about algorithmic biases; thus, individuals in the US tend to critically review algorithmic ethics. Korean users tend to accept algorithms without ethical question, assuming that the procedural aspects are legitimate, and they thus tend to trust algorithms more than US users. Korean users could consider AI to be more reliable and trustworthy than US users. However, they are probably more concerned with the outcome and performance of the algorithms in terms of whether they are accurate, predictable, and beneficial rather than whether the procedure is fair and just. Given the difference in procedure and performance, it can be said that algorithmic information processing is culturally sensitive and contextually defined. The relationship between heuristic and systematic processes is more intricate when applied to cross-cultural contexts. Contextual and cultural aspects of the algorithm are important, as people's attitudes toward AI are cultural and contextually dependent.

The algorithmic processing model in a chatbot context shows that interacting with algorithms takes in algorithmic sensemaking processes, wherein features of algorithms are cognitively processed to frame a heuristic of user motivation and to trigger user intentions for chatbot news. This sensemaking argument offers meaningful implications for the triadic relationship between algorithmic cues, trust, and credibility in chatbot services. The algorithm information processing perspective enables us to understand how users' trust is established, how it affects credibility, how algorithmic credibility functions, and how credibility is constructed (Van de Poel, 2020). This perspective is found to be a suitable frame insofar as the model argues that pre-behaviors and post-experiences influence user cognition, which in turn leads to satisfaction and intentions. Understanding the process by which algorithmic attributes lead to user decision can shape users' sensemaking of algorithms as well as how their actions influence their sensemaking. The model is suitable for algorithm systems because it is designed to examine user heuristics as a course of perceptions, experiences, and user formulation of trust based on cognitive processes.

4.6 HOW DO HUMANS PROCESS ALGORITHMIC INFORMATION?

Algorithmic information processing theory (Shin, 2021a) explains the stages of how users receive, perceive, and react to algorithms in their use of AI. Relevant research has consistently shown algorithmic information processing from a user perspective by integrating credibility into users' sensemaking processes (Hidalgo, 2021). The user understanding of algorithm-driven platforms is nonlinear and is not organized into structured, ready-made mechanical processes. How users feel, perceive, understand, and use algorithms depends on how they process algorithmic information. Users actively process the algorithmic decisions they receive from their cognitions and assess them in terms of fairness and transparency. Against rising concerns about algorithmic fairness and transparency (Sandvig et al., 2016), users seek to understand how algorithms work, how fake news operates, and how to protect against disinformation. Credibility dynamics offer insights into how credibility can be built and how it mediates the connection between functional and nonfunctional quality dimensions (Lee, 2018). Credibility in platforms is cognitively constructed in such a way that processes are transparently structured and understood in a human way (Guzman & Lewis, 2020). Established credibility allows users to believe that the recommendation is relevant and legitimate and that the source is credible.

Algorithmic information processes imply the existence of active (committed and conscious) roles for users to construct algorithms in AI platforms (Shin, 2021a). This role is consistent with the propositions of sensemaking that meaning is socially constructed through interaction (Dervin, 2003). The processes show how users create a shared understanding of their experiences with AI platforms. Some prior research has considered users to be passive recipients of recommended services that give their data to algorithms without consideration (Pu et al., 2012). With the rise of algorithm-driven technologies, the user's role has shifted from being a passive recipient of automated processes through media to a proactive creator of a preference profile that generates, adjusts, and modifies algorithms depending on the framing and contexts of their media consumption.

Users want to view what they would prefer to watch, they want to see what they would prefer to see, and their preferences become reinforced through the algorithmic process. The more users rely on algorithmic platforms, the narrower their perspectives become; this process has been described as an echo chamber or filter bubble (Sandvig et al., 2016).

Numerous relevant studies indicate that users are the creators of platforms as well as the source of algorithms by evoking profound subconscious cognitive processes (Lee, 2018). What users view through algorithms, in terms of their cognition, is a cognitively constructed representation that emulates the form of an accumulated experience shaped by a priori mental constructs. Algorithmic personalization has become a shared social reality that shapes daily lives and realities, affecting the perception of the outside (Shin & Park, 2019). Pursuant to their discussion, humans and algorithms are coevolving and creating reality together as they influence each other. Through credibility established by transparent fairness, humans and algorithms actively enhance each other's complementary roles.

4.7 HUMANIZING ALGORITHMIC INTELLIGENCE

Algorithmic credibility is based on the assumption that trust forms the foundation of our economies, societies, and sustainable development and that users, agencies, and societies will thus only ever be able to achieve the maximum promise of AI if credibility can be created in its design, development, and usage. Algorithmic credibility and trust should be key prerequisites of any AI system to prevent any harm from occurring and regenerating. AI practitioners systematically feed ethical codes and credibility principles into algorithms through periodic code and data audits.

In algorithmic platforms, users construct a sense of credibility with the processed information on fairness and transparency. Algorithmic credibility exists not in the algorithms but in the minds of users who perform sensemaking tasks. While various elements of a content source impact users' assessments of credibility, credibility lies in the process of the user of an algorithmic source. Affording robust user trust may assure users that their personal data is processed in compliance with transparent and acceptable manners, thereby engendering credibility for the recommendations and platforms and eventually leading to enhanced levels of user engagement. More engagement means greater algorithmic credibility to facilitate meaningful interactions.

Our discussion clarifies the sensemaking links among algorithmic attributes, algorithmic experiences, and users' interactions with algorithms.

Normative values provide users with cues for credibility, and established credibility enables users to use algorithms with interpretable and accountable assurance. User algorithmic credibility and trust processes open new areas for research. We offered an introductory conceptualization and fundamental operationalization of algorithmic information processing informed by the established, confirmed factors that influence algorithmic curation in AI adoption. Future work can examine in greater detail the underlying ties between credibility and algorithms and apply them to various AI contexts.

REFERENCES

Alexander, V., Blinder, C., & Zak, P. (2018). Why trust an algorithm? *Computers in Human Behavior, 89,* 279–288. doi:10.1016/j.chb.2018.07.026

Ananny, M., & Crawford, K. (2018). Seeing without knowing. *New Media and Society, 20*(3), 973–989. https://doi:10.1177/1461444816676645

Araujo, T. (2018). Living up to the chatbot hype: The influence of anthropomorphic design cues and communicative agency framing on conversational agent and company perceptions. *Computers in Human Behavior, 85,* 183–189. https://doi.org/10.1016/j.chb.2018.03.051

Bedi, P., & Vashisth, P. (2014). Empowering recommender systems using trust and argumentation. *Information Sciences, 279,* 569–586. doi:10.1016/j.ins.2014.04.012

Bishop, S. (2019). Managing visibility on YouTube through algorithmic gossip. *New Media & Society, 21*(11), 2589–2606. https://doi:10.1177/1461444819854731

Bohner, G., Moskowitz, G. B., & Chaiken, S. (1995). The interplay of heuristic and systematic processing of social information. *European Review of Social Psychology, 6*(1), 33–68.

Bolin, G., & Schwarz, J. (2015). Heuristics of the algorithm. *Big Data and Society, 2*(2), 1–12. doi:10.1177/2053951715608406

Borah, P. (2014). The hyperlinked world. *Journal of Computer-Mediated Communication, 19*(3), 576–590. https://doi.org/10.1111/jcc4.12060

Burgin, M. (1990). Generalized kolmogorov complexity and other dual complexity measures. *Cybernetics, 26*(4), 481–490. doi:10.1007/BF01068189. S2CID 121736453

Burrell, J. (2016). How the machine thinks. *Big Data & Society, 3*(1), 1–12. https://doi:10.1177/2053951715622512

Chawla, C. (2020). Trust in blockchains: Algorithmic and organizational. *Journal of Business Venturing Insights, 14,* e00203. https://doi.org/10.1016/j.jbvi.2020.e00203

Dervin, B. (2003). Sense-making's journey from metatheory to methodology to methods: An example using information seeking and use as research focus. In B. Dervin (Ed.), *Sense-making methodology reader* (pp. 141–146). New York: Hampton Press, Inc.

Diakopoulos, N. (2016). Accountability in algorithmic decision making. *Communications of ACM, 59*(2), 58–62. https://doi.org/10.1145/2844110

Dörr, K. N., & Hollnbuchner, K. (2017). Ethical challenges of algorithmic journalism. *Digital Journalism, 5*(4), 404–419.

Ford, H., & Hutchinson, J. (2019). Newsbots that mediate journalist and audience relationships. *Digital Journalism, 7*, 1013–1031. https://doi.org/10.1080/2167 0811.2019.1626752

Guzman, A., & Lewis, S. (2020). Artificial intelligence and communication: A human–machine communication research agenda. *New Media & Society, 22*(1), 70–86.

Hidalgo, C. (2021). *How humans judge machines.* Cambridge, MA: MIT Press.

Jones, B., & Jones, R. (2019). Public service chatbots: Automating conversation with BBC news. *Digital Journalism, 7*(80), 1032–1053. doi:10.1080/2167081 1.2019.1609371

Kim, D., & Lee, J. (2019). Designing an algorithm-driven text generation system for personalized and interactive news reading. *International Journal of Human-Computer Interaction, 35*(2), 109–121. https://doi.10.1080/10447318 .2018.1437864

Kitchin, R. (2017). Thinking critically about and researching algorithms. *Information Communication and Society, 20*(1), 1–16.

Klawitter, E., & Hargittai, E. (2018). It's like learning a whole other language. *International Journal of Communication, 12*, 3490–3510. http://doi.1932.8036/20180005

Kolkman, D. (2022). The (in)credibility of algorithmic models to non-experts. *Information, Communication & Society, 25*(1), 93–109. doi:10.1080/13691 18X.2020.1761860

Lee, M. (2018). Understanding perception of algorithmic decisions. *Big Data & Society, 5*(1), 1–16. https://doi:10.1177/2053951718756684

Lim, Y., & Heide, B. (2015). Evaluating the wisdom of strangers. *Journal of Computer-Mediated Communications, 20*(1), 67–82. https://doi.org/10.1111/jcc4.12093

Logg, J. M., Minson, J. A., & Moore, D. A. (2019). Algorithm appreciation: People prefer algorithmic to human judgment. *Organizational Behavior and Human Decision Processes, 151*, 90–103. https://doi.org/10.1016/j.obhdp.2018.12.005

Lokot, T., & Diakopoulos, N. (2015). News bots: Automating news and information dissemination on Twitter. *Digital Journalism, 4*, 682–699. doi.org/10.1080/21670811

Mahmud, H., Najmul Islam, A. K. M., Ishtiaque Ahmed, S., & Smolander, K. (2022). What influences algorithmic decision-making? A systematic literature review on algorithm aversion. *Technological Forecasting and Social Change, 175*, 121390. https://doi.org/10.1016/j.techfore.2021.121390

Obermeyer, Z., Powers, B., Vogeli, C., & Mulainathan, S. (2019). Dissecting racial bias in an algorithm used to manage the health of populations. *Science, 366*(6464), 447–453. doi:10.1126/science.aax2342

Oh, Y. J., Zhang, J., Fang, M. L. et al. (2021). A systematic review of artificial intelligence chatbots for promoting physical activity, healthy diet, and weight loss. *International Journal of Behavioral Nutrition and Physical Activity, 18*, 160. https://doi.org/10.1186/s12966-021-01224-6

Pu, P., Chen, L., & Hu, R. (2012). Evaluating recommender systems from the user's perspective: Survey of the state of the art. *User Modeling and User Adapted Interaction, 22*(4/5), 317–355.

Rai, A. (2020). Explainable AI. Journal of the Academy of Marketing Science, *48*, 137–141. doi:10.1007/s11747-019-00710-5

Reisdorf, B., & Blank, G. (2020). Algorithmic literacy and platform trust. In E. Hargittai (Ed.), *Handbook of digital inequality*. Northampton, MA: Edward Elgar Publishing.

Renijith, S., Sreekumar, A., & Jathavedan, M. (2020). An extensive study on the evolution of context-aware personalized travel recommender systems. *Information Processing & Management, 57*(1), 102078. https://doi.org/10.1016/j.ipm.2019.102078

Rietz, T., Benke, I., & Maedche, A. (2019). The impact of anthropomorphic and functional chatbot design features in enterprise collaboration systems on user acceptance. *Proceedings of the 14th International Conference on Wirtschaftsinformatik*. Siegen, Germany, February 24–27.

Rosenfeld, A., & Richardson, A. (2019). Explainability in human-agent systems. *Autonomous Agents and Multi-Agent Systems, 33*(6), 673–705. doi.org/10.1007/s10458-019-09408-y

Sandvig, C., Hamilton, K., Karahalios, K., & Langbort, C. (2016). When the algorithm itself is a racist: Diagnosing ethical harm in the basic components of software. *International Journal of Communication, 10*(2016), 4972–4990. 1932–8036/20160005

Shin, D. (2020). User perceptions of algorithmic decisions in the personalized AI system: Perceptual evaluation of fairness, accountability, transparency, and explainability. *Journal of Broadcasting & Electronic Media, 64*(4), 541–565. https://doi.org/10.1080/08838151.2020.1843357

Shin, D. (2021a). The perception of humanness in conversational journalism: An algorithmic information-processing perspective. *New Media & Society*. doi:10.1177/1461444821993801

Shin, D. (2021b). Why does explainability matter in news analytic systems? Proposing explainable analytic journalism. *Journalism Studies, 22*(8), 1047–1065. http://doi:10.1080/1461670X.2021.1916984

Shin, D. (2022). How do people judge the credibility of algorithmic sources? *AI and Society, 37*, 81–96. https://doi.org/10.1007/s00146-021-01158-4

Shin, D., Al-Imamy, S., & Hwang, Y. (2022). Cross-cultural differences in information processing of chatbot journalism: Chatbot news service as a cultural artifact. *Cross Cultural & Strategic Management*. https://doi.org/10.1108/CCSM-06-2020-0125

Shin, D., & Park, Y. (2019). Role of fairness, accountability, and transparency in algorithmic affordance. *Computers in Human Behavior, 98*, 277–284. doi:10.1016/j.chb.2019.04.019

Spiegelhalter, D. (2020). Should we trust algorithms? *Harvard Data Science Review, 2*(1). https://doi.org/10.1162/99608f92.cb91a35a

Sundar, S. (2020). Rise of machine agency. *Journal of Computer-Mediated Communication, 25*(1), 74–88. doi:10.1093/jcmc/zmz026

Swart, J. (2020). Tactics of algorithmic literacy: How young people understand and negotiate algorithmic news selection. *AoIR Papers of Internet Research, 2020.* https://doi.org/10.5210/spir.v2020i0.11342

Thurman, N., Moeller, J., Helberger, N., & Trilling, D. (2019). My friends, editors, algorithms, and I. *Digital Journalism, 7*(4), 447–469. https://doi.10.1080/216 70811.2018.1493936

Tsamados, A., Aggarwal, N., Cowls, J. et al. (2022). The ethics of algorithms: key problems and solutions. *AI & Society, 37,* 215–230. https://doi.org/10.1007/ s00146-021-01154-8

Van de Poel, I. (2020). Embedding values in artificial intelligence systems. *Minds and Machines, 30*(3), 385–409.

Wölker, A., & Powell, T. E. (2021). Algorithms in the newsroom? News readers' perceived credibility and selection of automated journalism. *Journalism, 22*(1), 86–103.

CHAPTER 5

Algorithmic Bias

H OW CAN WE ENSURE that AI systems are designed responsibly and produce effective outcomes? AI is as biased as humans are. Bias can originate from various venues, including, but not limited to, the design and unintended or unanticipated use of the algorithm or algorithmic decisions about the way data are coded, framed, filtered, or analyzed to train the machine learning. Algorithmic bias has been widely seen in advertising, content recommendations, and search engine results. Algorithmic prejudice has been found in cases from political campaign outcomes to the proliferation of fake news and misinformation. It has also surfaced in healthcare, education, and public service, aggravating existing societal, socioeconomic, and political biases. These algorithm-induced biases can exert negative impacts on a range of social interactions, ranging from unintended privacy infringements to solidifying societal biases of gender, race, ethnicity, and culture. The significance of the data used in training algorithms should not be underestimated. Humans should play a part in the datafication of algorithms.

5.1 WHY IS AI VULNERABLE TO BIAS?

Algorithms are human artifacts in that they are made, designed, trained, and applied by humans. Contrary to popular beliefs, AI is neither objective nor fair (Benjamins, 2021). An algorithm's performance largely hinges on the people who designed them, the code they used, the data they analyzed for the machine learning models, and the way they trained the models. As long as algorithms are human-made, they are naturally at risk of an inherent bias. Humans are under a cognitive bias, a pattern of deviation from rationality in judgment. This bias often leads to misrepresentation, wrong judgment, or mistaken interpretation. Humans create their own individual social reality more from their perceived world of reality and less from the objective input from it. Humans' cognitive biases are often transpired into algorithms they design, and thus, AI produces and amplifies the bias that humans entered. For AI to be aligned with human preferences, first, it must learn those preferences. AI systems that are trained on user behavior can mistake human biases for human values and then optimize for these biases. Learning human values by machine learning carries inherent risks. The underlying cause for AI bias lies in the conscious or unconscious human bigotry embedded in it throughout the development of algorithms. Hence, biases are quietly encoded during the process of algorithm creation.

Algorithms are programmed by people who – even with goodwill – can be prejudiced and discriminate within an unfair social world; thus, algorithms reflect and amplify the larger prejudices of reality (Shin et al., 2022). This type of amplification, which is called algorithmic amplification, is common in the platforms with which we interact every day because platforms such as Google, TikTok, Instagram, and YouTube are designed for organized data gathering, automated processing, distribution, and maximizing monetization of customer data. These platforms use vast amounts of data for algorithmic systems and have far expanded their capacities to drive people to decisions and behaviors that maximize monetization. The history of people's likes, clicks, comments, and retweets are the data that power the algorithmic amplification. By purposely ranking certain information higher, algorithms can amplify specific information while suppressing the representation of the rest. Some social communities benefit more from algorithmic amplification than others. Algorithmic amplification furthered some online content, becoming popular at the expense of other viewpoints. This is a reality on most of the platforms we use nowadays. The history of our likes, shares, and comments are the data driving the algorithmic amplification.

Herbert Simon (1957) presented the idea of bounded rationality, which is limited by imperfect information, cognitive ability, and time constraints. Bounded rationality plays a key part in the way humans design algorithms and in the way automation bias comes from. YouTube's recommendations amplify sensational content to increase the number of people on the screen as well as the duration for which they are on the screen. Platforms recommend videos concerning political information, inappropriate content, and hate speech as ways to maximize revenue. News recommender systems suggest either negative or incorrect information – likely to provoke rage, and fake news headlines are designed to be exaggerated news rather than real facts (Shin, 2021a). Programmers do the coding of the algorithms, markets choose the data used by the algorithms, and the algorithm designers decide how to apply the results of the algorithms. If the data analyzed by algorithms or used to train machine learning do not reflect the various parameters of users correctly, the results would be biased. Bias can enter into algorithms and machine learning because of preestablished social, cultural, and political inequalities in society and, thus, in people, which can impact decisions regarding how data is gathered, filtered, coded, or selectively analyzed to frame machine learning. It can be easy for people to let biases enter, which AI then algorithmizes and automates. That is why AI often makes decisions that are systematically unfair to certain groups of people. Algorithmic bias is a set of systematic faults in algorithmic systems that generate unfair discriminations, such as favoring a certain group of users over others (Fiske, 2022). A bias can be either intended or unintended, and it can emerge from a misunderstanding or a misinterpretation, that is, the intentional design of certain algorithms or unexpected decisions associated with the way data is gathered, analyzed, or included to train machine learning. Relevant inquiries have found that these biases can potentially cause significant harm to the public (Sundar et al., 2020). For example, facial recognition algorithms misjudge certain races more often than other people of color. A report published by the U.S. Department of Justice reported that facial recognition AI misjudges the races of colored people more often than noncolored ones. If used by law enforcement, facial recognition could run the risk of discriminatorily arresting people of color. Another study by the University of California Berkeley showed that some mortgage automations have charged colored people higher interest rates. Another study confirmed the finding that some bank loan algorithms have levied people of color to higher interest rates than white people. A study at Harvard University revealed that AI-driven speech recognition systems show significant racial disparities, with voice recognition misunderstanding 40% of words from minority users and only 11% of those from white users (Zarocostas, 2020). Also, the study found that virtual assistants with submissive Asian female voices reinforce stereotyped gender role bias. In 2021, it was revealed that Naver's recruitment system was biased against female candidates because the machine learning models were configured on datasets for weighing applications that reflected the male majority in the technology field. Algorithmic bias has commonly been found in social media platforms and search engine results. This bias can have serious effects on intensifying social stereotyping, biases, and prejudice. Algorithmic bias is prevalent in every aspect of our lives, as biased algorithms are embedded throughout healthcare, criminal justice, and employment systems, influencing critical decisions, operational work processes, and working rules. Machine learning helps technologies understand human rhetoric, bias, and discourse and has been found to reflect gender, racial, and class inequalities. Then, human biases, such as stereotyped sentiments attached to certain races, high-salary professions linked to a specific gender and race, and negative imaging of certain sexual orientations, become popularized in a wide variety of services. Indeed, algorithms can amplify human biases and societal stereotypes.

How do such biases enter into a set of algorithms? Humans write the codes in algorithms, select the data used by algorithms, and decide how to present the results of the algorithms. As humans develop an algorithmic structure, human biases inevitably are written into the algorithms. Biases are implanted through algorithmic data, machine learning embeds these biases, and AI reflects these biases in their performance (Huszar et al., 2021). Also, algorithms themselves contribute to biases. AI systems do not process and generate results only based on user data. They can also operate self-learning and self-programing algorithms based on secondary data, non-observational, and situational data such as synthesized data, simulations, bootstrapped data, or a combination of generalized assumptions or rules. Machine learning processes such data and learn from the data. Human rights may be negatively influenced by the use of such self-programming algorithms. People whose data is not processed or who have not otherwise been taken into consideration may also be directly involved and negatively impacted, particularly when algorithmic systems are used to inform critical decision-making. As Sartori and Theodorou (2022) state, "[Algorithms] are embedded within larger social systems and processes, inscribed with the rules, values and interests of a typically dominant group." Algorithms reveal glimpses of the existing structure of bias and inequalities that are embedded within our social, economic, and political systems. Without conscientious and rigorous mental investigations, it is easy for humans to intentionally and unintentionally input human biases into algorithms. Then, the biases are amplified, regenerated, and propagated. It is urgent and critical for AI researchers and industry leaders who develop AI technologies to test their algorithms to identify problems and potential biases.

Algorithmic bias can be seen across various platforms. Social media platforms that contain biased algorithms exacerbate misinformation, fake news, and disinformation. The 2020 U.S. presidential election result was partially influenced by algorithmic bias because platforms like Facebook and Instagram played as major biased news providers as they failed to filter out fake news. Also, fake news concerning COVID-19 has triggered panic, fear, and hatred in many places. There has been fake news accusing racial groups, illegal immigrants, and even governments of the diffusion of COVID-19. Certain political groups propagate fake news for the sake of political gains. Misinformation and disinformation about political campaigns harm democracy because people lose trust in the political system (Fiske, 2022). The real threat of algorithms is that they can amplify and magnify biases that already exist in the world. Machine learning models can encode and propagate at any scale the biases programmed intentionally or unintentionally. Realizing the seriousness of the bias issue, recent efforts have been made in legal frameworks like the Artificial Intelligence Act (2021) and the European Union's General Data Protection Regulation (2018). Most firms have started to run programs to fight against bias and societal inequalities. Amazon operates thorough antidiscrimination policies, recruits diverse races, trains to recognize potential employee bias, and promotes diversity. Regulators are pushing hard and tough drives. Legislation is underway in the state of California to require firms using AI in their business to examine AI for possible bias and report the audit results to authorities. However, all these efforts may be in vain if the AI models continue to operate in routine operations and the delivery of results remains inadvertently discriminating. Removing or mitigating algorithmic bias involves efforts beyond technical fixes. The tools and methods used to remove bias and reduce variance tend to cause another bias. Removing algorithmic bias should involve not only changing the algorithms or the systems but also changing cultural biases and social structures. Bias can perpetuate algorithmic inner systems as a result of preestablished social and cultural values. Society should continuously request that critical decisions be transparent, fair, and accountable, even as they become more and more algorithmized. If the self-regulatory model or ethical algorithm design does not work well properly, more systematic, ongoing, and legal measures auditing algorithms may be necessary.

5.2 TYPES OF ALGORITHMIC BIAS

The term "algorithmic bias" refers to systematic and repeatable faults in AI systems that engender unfair consequences (Fiske, 2022). When algorithmic systems are designed by private sectors, their foremost aim is to maximize profit out of users. Because machine learning in AI is capable of learning how users behave, it is capable of maneuvering users toward specific behaviors that are lucrative for firms. Potential threats from AI in terms of manipulating and maneuvering human behavior have been extensively researched. Researchers have identified several categories of bias in AI: measurement bias, selection bias, framing bias, confounding bias, and confirmation bias. These are the most common types of AI bias that creep into the algorithms. The forms of algorithmic bias are dependent upon where bias is coming from, the source of bias, and whether it is from humans, algorithms, and data (Zarocostas, 2020). Thus far, it has been indicated that the majority of biases result from humans because human prejudice causes and offsets other sources of bias. The types of algorithmic bias range from measurement bias to exclusion bias.

Measurement bias occurs because of underlying problems with the precision of the training data and how it is quantified and gauged. A model with flawed data collection methods or fallacious measurements leads to measurement bias and unfair results. For example, if an automatic survey system asks interviewees about preloaded questions, the responses will be inaccurate and biased due to systematic errors in the survey measurements. If an algorithm-driven recruiting system is programmed to select predetermined qualifications or schools, the system is biased toward specific criteria. A health care risk predictive algorithm is racially biased – because of its reliance on a faulty metric (systematically biased) for determining need.

Selection bias occurs when the data utilized in machine learning are not representative of the overall population and produce a distorted version of the real trend. This bias occurs in a database that has more cases of a particular type than others. If a database in algorithms contains overrepresented groups and underrepresented groups, which commonly happens because algorithms are trained to collect information online, the results would be biased. A chatbot technology trained only with the audio language data generated by males will have distorted voice recognition results when females interact with it. A face-recognition AI may train more faces of light-colored races than dark-colored races, thus producing biased performance in identifying darker-colored races. Biased sampling can lead to biased or invalid conclusions. This part of the sampling bias is a part of the overall selection bias.

The framing effect is a cognitive bias that influences human decision-making when said if different ways. People tend to believe in previously trustworthy cases and accept how the problem is framed and the information is presented; thus, the results obtained can vary and probably be biased. ProPublica, a nonprofit organization, found Correctional Offender Management Profiling for Alternative Sanctions (COMPAS) biased in predicting the possibility of recidivism. ProPublica alleged that the model used was biased against colored offenders, as the group was connected to a higher false-positive rate. ProPublica cited proof that the fairness metrics used in the COMPAS system breached the fairness criteria of equalized likelihoods and equality of opportunity. Most state and local governments now use algorithms to inform bail and parole decisions; we should demand that such decisions be transparent, fair, and accountable.

Confounding bias can result from the AI model if the model acquires incorrect associations by missing some information in the data or if it does not consider exhaustive information. This is a systematic misrepresentation in the measure of the relationship between exposure and the event caused by confounding the effect of that exposure of primary interest with irrelevant factors. This kind of bias stems from incorrect causalities that influence both inputs and outputs. For example, algorithms collect data on ice cream purchases and sunburns. Machine learning learns that higher ice cream buying is related to a higher chance of sunburn. Ice cream buying does not cause sunburn, and wrong causality between inputs and outputs is established, thus producing algorithmic bias.

Confirmation bias arises when the observer of the data experiment includes their envisaged results in the data. Bias can occur when humans begin with data research with biased thinking about their study, consciously or unconsciously. This bias commonly happens when people favor information that confirms previously existing thoughts or customs. There has been a misconception that left-handed people are more athletically effective or artistically creative than right-handed people. This would result in inaccurate data because this has never been proven, and algorithms can take this confirmation and turn it into bias.

Other forms of algorithmic bias include exclusion bias, recall bias, outliers, overfitting, underfitting, and reporting bias. Understanding potential types of biases in AI is the first step toward eradicating them. With that knowledge, it is critical to check the data analytic processes carefully to ensure that the datasets are as error-free as possible before they are put into the training stage.

5.3 A NEGATIVE FEEDBACK LOOP AND BIAS

A feedback loop is the part of a system in which some parts of the system's output are used as input for future operations (Shin et al., 2022). In AI, a feedback loop refers to the process of using the output of an AI system and corresponding user actions to train and reinforce models over time. The predictions and recommendations that are generated by AI are compared against the output, and feedback is provided to the model, making it learn from its errors. Feedback loops help AI systems learn what they did right or wrong, feeding them data that enables them to adjust their parameters to perform better in the future. This is a form of positive feedback loop that is sustained and supported by trust between humans and algorithms. A positive feedback loop is normally considered to have those components of a system that jointly increase each other's values when a stimulus occurs in one component. User profiles and recommendations shape a feedback loop. The users and the system are in a process of reciprocal confluence, where user profiles are continuously maintained and updated over time through interaction with the algorithm systems, and the quality of the algorithmic system is also improved by the user profiles.

Most AI applications in real-world systems follow this loop: data collection, data analysis, annotation and labeling, modeling, training, deployment, operationalization, use, feedback into the system, and use again. At every single point of this loop, bias can be added. Any model can spread biases in the training data. Humans should define their data sources and populations, their sampling methods for data collection, and the rules in the annotation and labeling task. The training model can proliferate and amplify the biases arising from prior analyses. The training model can enhance the feedback loop, which can contribute to rising bias. A relevant example of feedback loops producing bias can be found in Dressel and Farid's (2018) COMPAS study. It predicts the probability of an inmate recommitting a crime in the future. By analyzing the COMPAS algorithms, it was revealed that the algorithms are not accurate and are biased toward colored races. Non-White subjects were falsely labeled as having a high risk of committing a crime. This can be inferred by the fact that the recidivism model can produce different results with different parameters, such as gender, location, probability of family members who are convicts, and criminal history. With these parameters, the models are biased toward specific ethnicities and races. This finding is consistent with the ProPublica's investigative reporting that COMPAS is biased against colored defendants. Currently, COMPAS has been commonly criticized for propagating systematic racial bias.

Generally, certain popular items are suggested selectively, while the majority of other items are overlooked. These recommendations are then perceived by the users, and their reactions are logged and written into the system. This is called a negative feedback loop, which leads to generating and amplifying bias and misinformation. With negative feedback loops in place, bias causes further bias. Once a bias is included in the system, it can not only be compounded to generate discriminatory results that generate and amplify existing biased notions but also lead to algorithms developing their own ways to form biases. Because of the autonomous feature of negative feedback loops, algorithms can deviate from their intended goals, which can create significant obstacles when this bias mechanism goes unnoticed. Thus, designers and providers of algorithms should constantly monitor potential negative feedback loops that can lead to algorithms that generate biased decisions or even perpetuate gender and racial stereotypes.

5.4 FAKE NEWS, MISINFORMATION, AND AI

Hoaxes, misinformation, and disinformation are spread through online media. How do we fight against hoaxes and the spread of hoaxes? What is true, and what is a lie online? How do we detect fake news?

Fake news stories have proliferated online on social media platforms; this is in part because they are so easily and quickly shared with the advancement of algorithms. "Fake news" exists within a larger ecosystem of mis/disinformation by having different taxonomy; false information, misreporting, polarized content, satire, persuasive information, biased commentary, and citizen journalism (Molina et al., 2021). Fake news, misinformation, and, recently, deep fakes have been a persistent concern ever since the advent of AI. Fake news has already fanned blazes of distrust toward the media, politics, and established ideologies around the world (Huszar et al., 2021). Despite heightened awareness and public concerns, the problem persistently continues, and no single method appears to work to dissipate the problem. In fact, it has increased with the advancement of AI technologies. False information about COVID has thrived, which has had a significant impact on PCR tests and vaccination behaviors.

Misinformation poses a considerable challenge to public debate during political campaigns, and disinformation about Ukraine's War with Russia has led to confusion and anxiety. For the last several years, the volume and dissemination of fake news and misinformation have grown considerably (Zarocostas, 2020). There are several reasons for this exponential growth, including deepening the digital divide, economic disparity, decline in information literacy, existence of partisan media, and a decrease in public trust in media. Usually, while the wide use of social media is the main reason for the upsurge of misinformation, conventional media coverage can also contribute to the problem.

Constant exposure to the same misinformation makes it likely that someone will accept it. If AI detects misinformation and cuts the rate of its circulation, this can stop the cycle of reinforced information consumption and dissemination patterns. Machine learning algorithms can identify misinformation based on text nuance and the way stories are shared and distributed. However, AI detection of misinformation remains technically and operationally inaccurate. The contemporary algorithmic detection method is based on the evaluation of textual content. Although the method can determine the origin of the sources and the dissemination mechanism of misinformation, the essential limitation lies in how algorithms substantiate the actual nature of the information. Here are the real problems of fake news and misinformation: Fake news and misinformation are more about philosophical questions of how people deal with the truth than technical or mathematical questions of algorithms and AI. Many organizations now have fact-checker software tools for identifying fake news for reporters and journalists. Most fact-checking software performs basic functions, such as content-independent detection, by using tools that target the form of the content and by using deep fake identifying tools to check any manipulated content, image, and video. However, algorithms cannot check whether the content itself provides false information. This task should be done by human fact-checkers by searching for social media posts or online information with similar queries and information. Most current fact-checking approaches examine content by analyzing the semantic features of fake news. This approach may work at a basic level but faces bigger problems; for example, platforms like WeChat have language barriers, and fact-checkers cannot access the content of WhatsApp because it is an encrypted message. In addition, misinformation, in many cases, is an image that is difficult to investigate using textual techniques. If the amount of training data is sufficient, the AI-powered detecting model should be able to detect fake news and misinformation. However, the reality is that detecting such news demands prior social, cultural, and political knowledge, which AI algorithms still lack. We can prevent someone from making fake news and from spreading and promoting misinformation by applying AI-powered analytics that use anomaly detection, but eventually, AI cannot catch all problems related to fake news. Most current fake news detection systems require humans to work with AI to check the accuracy of information.

As the effort to counter fake news becomes more intensified, the industry will continue to work to find effective ways to sort out facts from fiction and to improve algorithmic tools that can help reduce disinformation. Many experts have argued that measures of fake news and misinformation should be contextualized within a broad understanding of digital ecology, taking into account societal and political factors (Lewandowsky et al., 2017). This implicates the need for a human-centered approach to fake news and misinformation. As much as fake news is related to human involvement in adoption, circulation, and dissemination, countering fake news should also be a human-centered approach to identifying fake news dissemination behavior, for example, by designing an explainable fake news spreader classifier based on the cognitive and psychological cues of users. The human-centered approach to fake news demands an understanding of the human aspects of fake news. Social and psychological user motivations play a key role in fake news generation and diffusion. Fake news spreading is underpinned by these motivations as well as by interactions among users and social contexts that enable them to write fake news. We cannot expect AI to solve the problems of misinformation and fake news because it is more than detecting fake news; it is a problem of trust and a lack of critical literacy. Beyond technological solutions, increasing people's media literacy through fact-checking and displaying the results can be a more effective solution. Combatting fake news, misinformation, and even deep fakes will require both AI and human efforts. While algorithmic tools are good at automatizing high quantities of information at rapid speeds, they lack the nuanced examination that human journalists can provide.

5.5 RESPONSIBLE AI

With a general definition of algorithmic bias, it is hard to define it as being systematically unfair. Thus far, there has been no widely accepted definition of the term systematic unfairness. Each firm has its own standard of fairness and its own measure of bias. Firms such as Microsoft, Google, and TikTok have their own policies. One problem with this is that the notion of responsible AI may be practiced unevenly across sectors (Mikalef et al., 2022). AI researchers and industry have proposed over 80 metrics, each of which defines bias by identifying how an algorithm considers different groups represented in a dataset differently. These groups are defined through a combination of sensitive characteristics (such as race, gender, age, sexual orientation, ethnicity, or religion). Deciding which bias metric is most effective requires a contextual interpretation of a use case. One place to start is with two general kinds of fairness: by representation or by error.

Despite active initiatives, addressing AI bias is challenging. As it is impossible to completely eradicate algorithmic bias, an alternative is considered: responsible AI. Now, society realizes the need for responsible AI, which is trustworthy, ethical, reliable, robust, well-governed, and understandable.

Responsible AI is a rising field of AI governance, and it is a blanket term that includes both legality and ethics. Responsible AI is the practice of designing, developing, and implementing AI with good intentions to empower people, and it fairly impacts users and society – allowing firms to generate trust and scale AI with confidence (Bastian et al., 2021). It serves as a reference framework that details how a particular organization addresses the risks around AI from both legal and ethical perspectives (Figure 5.1).

Responsible AI is the right thing to do, as it ensures that any AI system will be effective, follow regulations and norms, operate based on ethical values, and avert the potential for social and financial harm along the way (Mikalef et al., 2022). More and more government regulations are increasing their supervision of AI services and proposing legal frameworks that include guidelines and ethical standards. AI effect assessments and algorithmic auditing procedural frameworks are introduced to ensure that AI works responsibly and legitimately. An algorithmic audit is a key method in responsible AI that evaluates and rigorously frames fairness as well as systematically investigates how algorithms can be responsible by looking at the role they play in our society. Algorithmic audits operate on the assumption of the notion that algorithms are critically understood, framed, and regulated as they become fixtures of human life and social activities.

The European Media Freedom Act has called for news organizations to draft standards for the responsible use of AI in journalism. In the field of journalism, responsible AI is the design and implementation of algorithmic

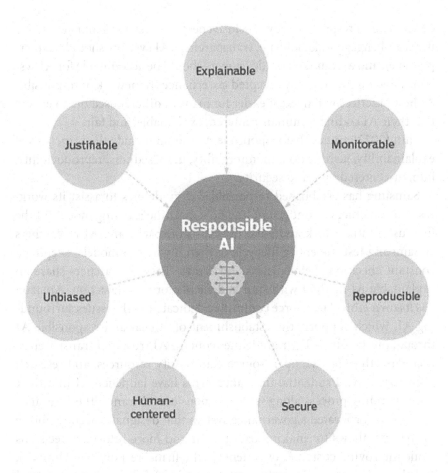

FIGURE 5.1   Components of responsible AI.

systems without "infringing on core values and human rights" (Dignum, 2019, p. 119). For example, Benjamins Field (2021) argued that news organizations should make generic technical choices regarding privacy, security, and safety that could lead to more responsible use of AI in the media. Currently, in most countries in the world, including the United States, there is no regulation of responsible AI. Despite this, leading industries, such as Microsoft, Meta, and Apple, have requested effective data and AI regulation.

Microsoft established its responsible AI principles in 2021. The guidelines propose six key principles in responsible AI: fairness (AI systems should treat all entities equitably), reliability and safety (AI systems should operate consistently and securely), privacy and security (AI systems should be secure and respect privacy), inclusiveness (AI systems should empower users and engage stakeholders), transparency (AI systems should be interpretable), and accountability (developers should be accountable for AI systems). Eventually, a widely accepted governance framework of responsible AI best practices will make it easier for firms in different sectors to ensure that their AI coding is human-centered, explainable, and fair.

Shin (2022) argues that responsible AI should include the principles of explainability, justification, monitorability, unbiasedness, reproducibility, human-centeredness, and security.

Samsung has established responsible AI guidelines to assist its workers and customers in understanding how machine learning models of the firm use routine work and what the AI's drawbacks are. AI researchers at Samsung test the entire lifecycle of their algorithm models, keeping a constant check on their fairness and errors. These researchers share an explainable AI method with staff and clients for bias detection. Amazon has its own ethical task force committee dedicated to the issues surrounding AI, which supports the establishment of ethical and responsible AI throughout the firm. Their guidelines contain AI trust and transparency, everyday ethics for AI, open-source community resources, and research into trusted AI. Prudential insurance firms have launched AI into their claim-handing process along with a responsible AI frame. The insurance firm has incorporated a governance system that designates accountability for AI and allows for making transparent and more informed decisions while improving customer experiences. The firms' responsible AI model, called the Dynamic Property Assessment Model, helps predict a total financial loss with a high level of confidence while enabling transparency regarding the tool for processing insurance claims.

A responsible AI framework incorporates details on the kind of data that can be gathered and analyzed, the way models can be assessed, and the best way to implement and supervise models. The framework can also define who is responsible for any harmful AI results. The goal of responsible AI is to design understandable, transparent, and socially acceptable AI systems. The most important component of the framework is explainability. When we interpret models, we discover an explanation of how they generate personalization. An AI system can diagnose patients with cancer, and an algorithmic model can deny loan applications. Users would require an explanation regardless of the correctness of the decisions. Responsible AI can describe how we can design understandable models or when it is appropriate to use one that is less understandable.

The term fairness is related to explainability. While there is no consensus on a single definition, it is known that fairness in AI should treat all people fairly. It is easy for AI systems to make decisions that victimize underserved groups of people, which results from bias in the data used to train machine learning. The more explainable the model, the easier it is to warrant fairness and detect possible biases. There is a need for a responsible AI framework to define how people perceive fairness and what to do when algorithms are found to create biased predictions.

Eventually, trust is the most significant factor in responsible AI. If users trust AI, they will continue to use the services. People do not trust AI systems that use data that we are unsure about sharing (such as sensitive profile data of race, sexual orientation, or political beliefs), considering that such systems produce biased results. People do not trust AI if they believe it may harm them. Explanations for decisions and accountability for those decisions go a long way toward building this trust. This need for trust drives self-regulation among companies that use AI. In 2022, the European Commission announced the ethics guidelines for trustworthy AI – the core principles that AI should have to be deemed responsible and faithful. These guidelines would help firms ensure that their AI systems meet the same standards. However, a more important question is: Can we trust that firms can regulate themselves?

5.6 FAIRNESS AND TRANSPARENCY IN ALGORITHMS

Fairness and transparency are becoming important considerations in the use of algorithms for the recommendation and delivery of digital content (Ananny & Crawford, 2018). Automated data gathering and sharing may involve processes that are unfair, flawed, opaque, or unaccountable (Crain, 2018). Over-the-top platform content recommendations embody these issues in highly visible applications. Fairness and transparency bring up vital prerequisites in the design and development of algorithm-supported platforms (Kitchin, 2017), which are purportedly designed to offer accurate and reliable recommendations for users (Lepri et al., 2018). It remains unresolved whether such recommendations match user interest, how the analytic processes are done, and whether the outcomes are legally responsible (Diakopoulos & Koliska, 2016). Thus, fairness and transparency emerge as fundamental requirements in the use of algorithms on media platforms (Montal & Reich, 2017). When transparent, open, and fair services are provided, users are more likely to consider that the recommendations are of high quality (Shin & Park, 2019). Highly transparent platforms can grant users a sense of personalization, as responsible and fair recommendations afford users a sense of trust that promotes satisfaction and a willingness to continue using them and subscribe to them (Soffer, 2019). Open visibility and clear transparency of relevant recommendations boost the user interpretability of the system and search performance (Shin, 2020).

AI incessantly affects the everyday lives of billions of media users (Moller et al., 2018). Algorithms are widespread and accepted in practice, but their popularity comes at the expense of limited transparency, systematic prejudice, and nebulous responsibility (Wölker & Powell, 2021). Algorithmic filtering procedures may lead to more impartial, and thus possibly fairer, processes than those processed by humans. Yet, algorithmic recommendation processes have been criticized for their bias to intensify/reproduce prejudice, information asymmetry, distortion of facts, and the black box process of decision-making (Sandvig et al., 2016). Algorithmic bias may increase algorithmic inequality in that machine learning automates and propagates unjust and discriminatory patterns (Hoffmann, 2019). Recent over-the-top (OTT) platforms (Hulu, Disney, or Netflix) have faced similar problems (Shin, 2020). While these platforms offer personalized and relevant content in innovative and interactive manners, ethical and privacy issues remain intertwined with algorithmic personalization in a complicated way (Araujo et al., 2022). Hence, the issues regarding how to safeguard the goals, values, and personalizing processes of platforms; to what extent users need to share personal information with algorithms; and how to balance privacy and algorithmic personalization remain controversial (Hoffmann, 2019). Underlying these questions are concerns about how to mitigate bias and discrimination in data and the urgent need to understand how to design algorithmic systems that are transparent and fair (Crain, 2018). As ethical concerns have peaked recently with the rise of OTT algorithms, the opacity of black box algorithm processes has led to calls for investigation on transparency and fairness (Sandvig et al., 2016).

Recent research (e.g., Shin & Park, 2019; Dörr & Hollnbuchner, 2017) has highlighted the normative implications and problems of these algorithms, which are summarized by fairness, accountability, and transparency (FAccT). In particular, transparency and fairness of algorithmic systems while processing user-sensitive data emerge as key attributes for those algorithmic systems to be trustworthy (Shin et al., 2022). This issue will be even more important when OTT increasingly relies on algorithms, and people rely more on algorithms than social influence when making judgments. AI has become pervasive across all media industries and service functions. A key question arises regarding how to govern these OTT algorithms effectively and legitimately while ensuring that they are user-centered and socially responsible. Despite their significance, few studies have researched normative values in OTT contexts. As AI becomes even more prominent in media platforms, the vital issues to consider will be how users can make sense of fairness in their algorithm use, how users understand algorithmic transparency, and how they accept algorithm-based processes (Thurman et al., 2019). As these normative concerns have given rise to the demand for a better explanatory framework that effectively addresses them (Graefe et al., 2018), we discuss these concerns from a user's perspective: how fairness and transparency influence the form and content of sensemaking processes, and we do this with a focus on how users formulate credibility through quality evaluation. User experiences may be based on subjective standards of fairness and transparency, generating tension between uniform solutions deemed to conform to ethical requirements and potentially diverging from user experiences. However, there is little understanding of how fairness/transparency has been conceptualized and how users experience such factors. There is thus a gap between transparency in practice and the goal of fairness since ethical standards predominantly serve internal stakeholders, reflecting algorithm providers' interests rather than users' interests and benefits. For example, numerous explainable AI cues (such as counterfactual explanations, feature importance scores, or source information checking) are not for users who are actually affected by the algorithms but rather for AI programmers who use explainability to debug the algorithm itself.

5.7 THEORIZING THE EFFECTS OF FAIRNESS AND TRANSPARENCY ON SENSEMAKING PROCESSES

An OTT platform is a media service offered directly to viewers via the internet that delivers content over the top of another platform. Contemporary OTT platforms, such as Amazon Prime, Netflix, YouTube Premium, and Naver, have significantly transformed the media landscape by utilizing the power of AI. OTT platforms provide innovative user experiences as well as disruptive and scalable business models (Shin et al., 2022). On one hand, these platforms provide access to millions (on entertainment platforms with professionally produced content) or billions (on social media and user-generated content platforms) of pieces of digital content to users in a user-customized manner through algorithmically driven recommendations. On the other hand, suppliers can reach millions of users in a direct and scalable way without dealing with multiple instances of gatekeepers (Parker et al., 2016). OTT platforms can be seen as high-value opportunities that generate further value through every user interaction. A consequence of this intrinsic, self-propagating value creation is the emergence of monopolies, as competitors are shut out, and the roles of different market participants and intermediaries are absorbed under the single interest of maximizing the efficiency and profit of the platform. These effects intensify if the platform itself is also the producer of the digital goods offered.

In OTT markets where algorithms have been widely embedded and operated on, the question of fairness and transparency becomes relevant. Fairness and transparency emerge as fundamental attributes against the growing adoption of machine learning algorithms to process large amounts of user data. Algorithms embedded in OTT platforms can replicate possible biases encoded in the data. As these platforms dominate markets and reduce user agency, a societal responsibility beyond the profitability of a single enterprise must be stipulated, making it imperative that fairness/transparency be followed in their conduct. While governments and practitioners have examined the legal and ethical concerns that arise from the widespread diffusion of algorithm implementation, little work has attempted to explain how we can define transparency/fairness in algorithms and how fair and transparent algorithms can be embedded and represented in an operational algorithmic system (Shin, 2021b).

5.8 FAIRNESS AND TRANSPARENCY GROUNDED IN USERS' PERSPECTIVES: TRANSPARENT FAIRNESS

Producing user-centered algorithms is dependent on developing algorithmic platforms that use more responsible and transparent processes. Relevant recent studies have shown that general perceptions of accuracy and transparency are not solely objective responses to media content (Shin et al., 2022). Rather, our discussion lends strong support to the ongoing argument that, to a significant extent, perceived transparency and fairness in algorithmic platforms are subjective, similar to perceptions of information in general (Kemper & Kolkman, 2019). Thus, transparency and fairness can be considered perceptions held by users, which are more subjective than based on objective criteria (Shin & Park, 2019). There are numerous dimensions by which we can measure how "transparent and fair" a recommendation is. However, in reality, it is challenging to measure transparency and fairness directly. This methodological difficulty seems to be partially contributing to the argument: transparency and fairness may depend on users' perceptions and understanding. Per the sensemaking proposition, the meanings are socially constructed, negotiated, and communicated through various interactions within the user's mind and cognitively reenacted within users' sensemaking processes (Gu et al., 2021). Rather than such issues being unvaryingly or irreversibly provided to users, users construct their own versions of transparency and fairness based on their levels of existing trust and other intrinsic personal factors (Lepri et al., 2018). This proposition provides a reference point for sensemaking theory in algorithmic contexts. Transparency and fairness are cognitively and contextually formed realities of the user and depend on that person's perceptions.

This point guides toward some more advanced insights that algorithmic consumption and interactions are significantly associated with users' sensemaking processes, which are significantly interwoven with users' perceptions of transparency and fairness. Based on the closely interwoven nature of transparency and fairness, relevant works have proposed the confounding idea of transparent fairness (Shin et al., 2022). Recent research confirms that the two are essentially related, and even more so in the case of algorithms. When an algorithm is fair, people consider it transparent. Conversely, when an algorithm is transparent, people consider it fair. A transparent system affords the notion of fairness. Transparency remains relevant to fairness, but it can neither be necessary nor sufficient for it. Combining transparency with ideas of fairness, we propose requirements for OTT media platforms. While previous studies have extensively examined these two factors, little is known about how they are related conceptually and operationally. Our discussions provide a relevant theoretical contribution by clarifying the interrelated nature of algorithmic bias and thus proposing a new algorithmic attribute: transparent fairness. Conceptually, transparency is fundamentally linked to fairness, constituting two sides of a coin, working both technologically and emotionally to build the credibility of media-associated algorithms. Transparency can be considered a system-related attribute, whereas fairness can be viewed as a user-related attribute. Transparency is a technical attribute that algorithm systems bear and involve, whereas fairness is an emotional judgment of how users feel when transparency is confirmed. Conversely, when fairness is ensured in the system, transparency increases, as it encompasses the notion of fairness. This conceptual contribution can be heuristically meaningful in terms of redefining such concepts in the context of OTT platforms. However, since the two attributes have not been fully explicated, we do not know the extent to which such attributes influence or are influenced by the quality of algorithmic performance; hence, whether a system is fair and transparent, to what extent the two overlap, and how the two are interrelated remains blurry.

Transparent fairness is an extended concept of fairness with respect to the algorithm's ability to provide explainable features and interpretable mechanisms. This extended concept of fairness is necessary in cases where fairness is prioritized, such as in OTT platforms, where data are labeled with previous user data. Some people consider algorithms fair, even if the system is not transparent. Some people perceive algorithms as transparent, despite the perception that the system is unfair and biased. The concepts of fairness and transparency combine to yield a few possibilities: transparent and fair, opaque and fair, transparent and unfair, and opaque and unfair. Transparent fairness goes beyond procedural fairness, where the latter is concerned with the automated procedures used by algorithms, giving users a sense of absolute and complete fairness regarding proper data analytical procedures and recommendation outcomes. Transparent fairness can be an underlying theoretical attribute of algorithms in OTT, as well as a practical scale for evaluating the performance and results of algorithms.

Against the rising trend of algorithmic bias, there are challenges in addressing AI bias. Increasing behavioral manipulation via AI calls for policies that warrant human control and user autonomy in AI. For example, the European Commission's Ethics Guidelines for Trustworthy AI states user control over AI in order to make sure AI does not deceive, manipulate, or mislead. Yet, the policies remain still nascent. With the lack of standards to follow, firms must, by themselves, figure out what kind of information would cause such biases in the algorithms they use. The continuous evaluation of data and different legal guidelines also makes it difficult to control algorithmic bias. AI models have to access new data and work with historical data. The set of standards and rules that has been used in past years will most likely change; hence, the results may be unexpected. These difficulties raise the need for algorithmic literacy and awareness for users. The best way to reduce AI bias and increase algorithmic trust is to increase awareness in people that they must be AI literate.

REFERENCES

Ananny, M., & Crawford, K. (2018). Seeing without knowing. *New Media & Society, 20*(3), 973–989. doi:10.1177/1461444816676645

Araujo, T., Helberger, N., Kruikemeier, S., & de Vreese, C. H. (2020). In AI we trust? Perceptions about automated decision-making by artificial intelligence. *AI & Society, 35*(3), 611–623. https://doi.org/10.1007/s00146-019-00931-w

Bastian, M., Helberger, N., & Makhortykh, M. (2021). Safeguarding the journalistic DNA: Attitudes towards the role of professional values in algorithmic news recommender designs. *Digital Journalism, 9*(6), 835–863. doi:10.1080 /21670811.2021.1912622

Benjamins, R. (2021). A choices framework for the responsible use of AI. *AI and Ethics, 1*(1), 49–53. https://doi.org/10.1007/s43681-020-00012-5

Crain, M. (2018). The limits of transparency. *New Media & Society, 20*(1), 88–104. doi:10.1177/1461444816657096

Diakopoulos, N., & Koliska, M. (2016). Algorithmic transparency in the news media. *Digital Journalism, 5*(7), 809–828. https://doi.org/10.1080/21670811 .2016.1208053

Dignum, V. (2019). *Responsible artificial intelligence: How to develop and use AI in a responsible way*. London: Springer Nature.

Dörr, K., & Hollnbuchner, K. (2017). Ethical challenges of algorithmic journalism. *Digital Journalism, 5*(4), 404–419.

Dressel, J., & Farid, H. (2018). The accuracy, fairness, and limits of predicting recidivism. *Science Advance, 4*(1). doi:10.1126/sciadv.aao5580

Fiske, S. (2022). Twitter manipulates your feed: Ethical considerations. *Proceedings of the National Academy of Sciences, 119*(1). doi:10.1073/pnas.2119924119

Graefe, A., Haim, M., Haarmann, B., & Brosius, H.-B. (2018). Readers' perception of computer-generated news: Credibility, expertise, and readability. *Journalism, 19*(5), 595–610. https://doi.org/10.1177/1464884916641269

Gu, J., Yan, N., & Rzeszotarski, J. (2021). Understanding user sensemaking in machine learning fairness assessment systems. *Proceedings of the Web Conference*, ACM, New York. https://doi.org/10.1145/3442381.3450092

Hoffmann, A. L. (2019). Where fairness fails: Data, algorithms, and the limits of antidiscrimination discourse. *Information, Communication & Society, 22*(7), 900–915. doi:10.1080/1369118X.2019.1573912

Huszar, F., et al. (2021). Algorithmic amplification of politics on Twitter. *Proceedings of the National Academy of Sciences, 119*(1). doi:10.1073/pnas.2119924119

Kemper, J., & Kolkman, D. (2019). Transparent to whom? *Information, Communication & Society, 22*(14), 2081–2096. https://doi.org/10.1080/13691 18X.2018.1477967

Kitchin, R. (2017). Thinking critically about and researching algorithms. *Information, Communication & Society, 20*(1), 14–29. https://doi.org/10.1080/136 9118X.2016.1154087

Lepri, B., et al. (2018). Fair, transparent, and accountable algorithmic decision-making processes. *Philosophy & Technology, 31*(4), 611–627. https://doi:hdl. handle.net/1721.1/122933

Lewandowsky, S., Ecker, U. K., & Cook, J. (2017). Beyond misinformation: Understanding and coping with the post-truth era. *Journal of Applied Research in Memory and Cognition*, 6(4), 353–369. https://doi.org/10.1016/j.jarmac.2017.07.008

Mikalef, P., Conboy, K., Lundstrom, J., & Popovič, A. (2022). Thinking responsibly about responsible AI and the dark side of AI. *European Journal of Information Systems*. https://doi.org/10.1080/0960085X.2022.2026621

Molina, M., Sundar, S., Le, T., & Lee, D. (2021). Fake news is not simply false information. *American Behavioral Scientist*, 65(2), 180–212. doi:10.1177/0002764219878224

Moller, J., Trilling, D., Helberger, N., & van Es, B. (2018). Do not blame it on the algorithm: An empirical assessment of multiple recommender systems and their impact on content diversity. *Information, Communication & Society*, 21(7), 959–977. doi:10.1080/1369118X.2018.1444076

Montal, T., & Reich, Z. (2017). I, robot. you, journalist. who is the author? *Digital Journalism*, 5(7), 829–849. doi:10.1080/21670811.2016.1209083

Parker, G., Van Alstyne, M., & Choudary, S. (2016). *Platform revolution*. New York: WW Norton and Company.

Sandvig, C., Hamilton, K., Karaholios, K., & Langbort, C. (2016). When the algorithm itself is a racist. *International Journal of Communication*, 10, 4972–4990.

Sartori, L., & Theodorou, A. (2022). A sociotechnical perspective for the future of AI: Narratives, inequalities, and human control. *Ethics and Information Technology*, 24(4), 1–11. https://doi.org/10.1007/s10676-022-09624-3

Shin, D. (2020). User perceptions of algorithmic decisions in the personalized AI system: Perceptual evaluation of fairness, accountability, transparency, and explainability. *Journal of Broadcasting & Electronic Media*, 64(4), 541–565. https://doi.org/10.1080/08838151.2020.1843357

Shin, D. (2021a). The perception of humanness in conversational journalism. *New Media and Society*. http://doi:10.1177/1461444821993801

Shin, D. (2021b). Embodying algorithms, enactive AI, and the extended cognition: You can see as much as you know about algorithm. *Journal of Information Science*. http://doi:10.1177/0165551520985495

Shin, D. (2022). How do people judge the credibility of algorithmic sources? *AI and Society*, 37, 81–96. https://doi.org/10.1007/s00146-021-01158-4

Shin, D., & Park, Y. (2019). Role of fairness, accountability, and transparency in algorithmic affordance. *Computers in Human Behavior*, 98, 277–284. doi:10.1016/j.chb.2019.04.019

Shin, D., Zaid, B., Biocca, F., & Rasul, A. (2022). In platforms we trust? Unlocking the black-box of news algorithms through interpretable AI. *Journal of Broadcasting and Electronic Media*. https://doi.org/10.1080/08838151.2022.2057984

Simon, H. (1957). *Administrative behavior: A study of decision-making processes in administrative organization*. 2nd edition. New York: Macmillan.

Soffer, O. (2019). Algorithmic personalization and the two-step flow of communication. *Communication Theory*. https://doi.org/1093/ct/qtz008

Sundar, S., Kim, J., Beth-Oliver, M., & Molina, M. (2020). Online privacy heuristics that predict information disclosure. *CHI '20*, April 25–30. https://doi.org/10.1145/3313831.3376854

Thurman, N., Moeller, J., Helberger, N., & Trilling, D. (2019). My friends, editors, algorithms, and I. *Digital Journalism*, 7(4), 447–469. https://doi.org/10.1080/21670811.2018.1493936

Wölker, A., & Powell, T. (2021). Algorithms in the newsroom? *Journalism*, 22(1), 86–103. https://doi.org/10.1177/1464884918757072

Zarocostas, J. (2020). How to fight an infodemic. *The Lancet*, 395(10225), 676.

CHAPTER 6

Explainable Algorithms

Human–AI interactions face challenging issues, such as a lack of transparency regarding how algorithmic results are produced and an absence of reciprocally co-constructed interaction. It has become a norm that the AI processes should be understood and explainable in order for AI systems to be trusted. It is becoming almost a norm to explain why algorithms make a decision in a particular case, as the importance of explainability has made us reconsider using AI, particularly in critical domains, such as health, public service, and public safety. As AI faces a transparency crisis, explainable AI is considered an alternative solution to these transparency problems by ensuring transparency so that users can understand the internal processes of algorithmic models. When AI incorporates the feature of explainability, the services provide an in-depth understanding of the algorithms and their generated results. We further argue for human-interpretable explanations by discussing the dimensions and effects of interpretability and understandability in AI on user attitudes and heuristics. Designing more interpretable AI can be a meaningful step toward developing an ethical and societal discourse around algorithms that can open their opaque black box.

6.1 WHY EXPLAIN? EXPLAINING EXPLAINABILITY

Debates on the need for explainable AI have been triggered by discussions on algorithmic transparency and fairness (Shin, 2021). As more and more organizations embed AI and algorithms within their organizational processes and automate decisions, transparency is increasingly needed in how these processes make decisions. Questions arise as to how we can trust AI if we do not know what it is and what it does. How do we establish trust in AI, and how do we create transparency while maximizing the efficiencies that algorithms bring? Per Bucher's (2018, p. 60) argument, "While we cannot ask the algorithm in the same way we may ask humans about their beliefs and values, we may indeed attempt to find other ways of making it speak." Explainable AI is proposed as one of the ways to make algorithms speak and solve problems.

While there are various definitions, explainability refers to the translation of technical logic and decision results into understandable, comprehensible structures that are appropriate for making assessments (Arrieta, 2020; Shin, 2021). Explainability affords users actionable insights into what they should do in terms of future interactions with algorithm systems. Explainable AI is defined as algorithms and machine learning that can present comprehensible human reasoning for their results or processes (Rai, 2020). Explainability is essential for establishing faith, trust, and rapport between AIs and their users. The need to understand algorithmic decision-making is an increasing concern, especially when users receive unexpected results and undesirable consequences. Many users have begun to ask questions about the usefulness of algorithms if we are unable to examine them and understand them.

In journalism, explainable systems can be applied to reporting, as exemplified by explanatory journalism (Shin, 2021). Explainable AI helps newsrooms to present ongoing news stories in a more understandable and accessible way by offering clear explanations on chosen issues. Explainability allows algorithmic journalism developers, reporters, and readers to better understand why certain news recommendations are suggested and to improve them as needed. It is not about explaining the news but rather about explaining why the reader receives the news. This is critical in the context of journalistic value in AI, as it enables journalism to identify potential biases and discrimination against certain values. On the reader's side, when explainable (thus understandable) AI is open to direct questioning – and, further, if the algorithmic journalism itself is open source – a product can be examined line by line. Explainability in algorithmic journalism means that users grasp the algorithmic journalism process, and algorithmic journalism is encoded to explain its goals, logic, and recommending processes in ways that users can recognize. A lack of explainability can be ethically problematic in algorithmic journalism, as it generates ignorance on the part of users who read AI-generated news (Rosenfeld & Richardson, 2019).

As AI becomes more advanced and sophisticated, explainability becomes increasingly important (Chazette & Schneider, 2020). Users are challenged to know how the algorithms produce results. The entire inferring process is turned into what is commonly referred to as a black box that is hard to comprehend. A stack of data continually adds to the level of the black box, and often even the designer or scientists who design the algorithms cannot retrace or understand how the machine learning produced certain results because they cannot know what is occurring inside them. In reality, opening up the black box of algorithms is practically impossible as algorithms are protected and remain hidden in the name of trade secrets and as a right of intellectual property. For instance, looking inside TikTok's algorithms is nearly impossible because, first, the platform would not allow auditing, and second, their implementation and deployment constantly change. The practice of algorithms and deep neural networks behind Netflix's recommender system is proprietary, so to a large extent, they are kept purposely opaque to ensure the firm's financial advantages and copyright over the algorithms. Research has suggested new ways for users to understand algorithms through enhanced measures for fairness and transparency, including disclosure requirements, regulatory control, and explainability (Combs et al., 2020).

There have been extensive research efforts on explainability and related concepts, such as transparency, in various AI contexts (Holzinger, 2016). Shin (2020a) conceptualized algorithmic explainability as information associated with a certain decision and why the decision was made. It is a way to increase transparency about the data that were analyzed, how the data were selected, how the data were trained, and how the data are implemented in relation to the decision. While emphasizing the importance of meaningful explanations, Shin (2021) indicated that current explainable AI remains basic and that most explanations do not make sense, or at least are not relevant to what users really wish to understand. Criticism of the limited number of current explanations has increased, thus raising the pressure to produce appropriate and accurate explanations about the process and logic of algorithmic outputs (Vallverdú, 2020). This criticism has led to a shift in focus away from algorithms and toward users in terms of how users interpret such explanations, how they reason causality and causal inference (Combs et al., 2020), and through what process they work to understand the issues in algorithms that are vague and unclear (Ehsan & Riedl, 2019). These studies are based on common assumptions that explainability is much more than outlining technical concepts; rather, it is about empowering users to utilize the information provided to them to challenge algorithmized decisions to ensure their basic right to know.

There are numerous advantages to knowing how an algorithm-enabled system has led to certain decisions (Rai, 2020). In credit score systems, for example, explainable AI can improve the credit score model used by financial institutions. In a traditional model of credit scores, it is nearly impossible to know how and why each input influences the score. With explainable AI, banks now show the feature percentage influence of each input on the output. Such explainability allows model designers, banks, regulators, and customers to understand why and how certain scores are generated and to correct them as needed. This knowledge is critical in the context of bias and the integrity of algorithms because it will enable firms to detect latent discrimination against certain social groups and ethnic minorities. As well as establishing trust and credibility and adopting good practices around responsibility and ethics, significant benefits can be obtained from being on the front foot and establishing explainability. Explainability can also assist developers in ensuring that the system is operating as it should be. Aside from being necessary to comply with regulatory requirements, it is also helpful for responding to those affected by algorithmic decisions. Articles 13 and 22 of the European Union's GDPR, for example, state that when citizens are affected by decisions made through algorithmic processing, they are entitled to a reasonable explanation of the logic involved. In addition, the 2020 California Consumer Privacy Act (CCPA) states that citizens have a right to know the reasonings made about them by AI systems and what data were employed to make those reasonings. As legal demand escalates for transparency, academia and industry push explainable AI forward to meet new requirements and ever-growing user expectations. As AI develops, more regulations need to be launched to ensure the transparent and explainable implementation of AI.

Despite the huge potential and opportunities, there are concerns regarding the extent to which human decision-making will be replaced by algorithms (Dörr & Hollnbuchner, 2017). Concern is rising over the transparency of algorithm services, which requires firms to be truthful regarding the goal, service, and essential procedures of algorithms used to search for, process, and deliver information. The issues of fairness, transparency, and accountability can significantly undermine algorithms and AI services by generating a series of undesired and even hazardous glitches in AI systems (Ferrario et al., 2020). Explainable recommendations, which accompany explanations about why an item is recommended, networks used in RS are autoencoder [22–25], convolutional neural network (CNN) [26–29], recurrent neural network (RNN) [30], [31], generative adversarial network [32–34], graph neural network [35], [36].

Deep neural networks are increasingly being used in recommender systems to deal with more complex scenarios, such as dynamic environments, various data sources, and diverse data representations. They are developing methodologies and building models using a variety of deep neural networks in order to better understand the preferences of users.

5.4.3 EVOLUTIONARY COMPUTING IN RECOMMENDER SYSTEMS

When treating recommendation as a multi-objective optimization problem, evolutionary algorithms (EAs) are used to combine the outputs of multiple recommendation algorithms. Additionally, they are used to generate user/item profiles and to manage recommendation ratings. The application of EAs in recommender systems can be broadly divided into the following three categories.

Zhang and Yeung (2010) propose a new probabilistic multi-objective evolutionary algorithm with a new crossover operator called a multiparent probability genetic operator and a new topic diversity indicator that strikes a good balance between accuracy and variety. Karabadji et al. (2018) improved a memory-based CF method by using multi-objective optimization to find neighbours in order to achieve accurate and diverse recommendations. EA is also applied in user/item profile optimization as can be seen in (Chen et al. 2017; Rana and Jain 2015). EA has also been applied to rating optimization. Multi-criteria ratings can be integrated into recommender systems, according to [41–43]. This type of algorithm uses multi-criteria ratings to make recommendations that take into account more complex preferences from the user.

5.5   CASE STUDIES

5.5.1   NETFLIX

Apparently, the value of a Netflix recommender system can be measured by the increase in member retention. Although the development of personalization and recommendation technologies has resulted in significant enhancements in user retention, these enhancements have not been as substantial as expected. The primary decision that is made by the recommender system is to choose which videos to show each subscriber on their Netflix homepage after they have logged into their profile on any device. This choice can be made at any time This personalization task is subdivided into a number of smaller subtasks using recommender systems that are designed to accommodate a wide range of member requirements. Each of these subtasks may be governed by a distinct algorithm [44].

Instead of a single model driving all Netflix recommendations, the company uses a set of techniques that all work together to achieve the same goal: customer satisfaction. The Netflix Recommendation System is the name given to this collection of methods. Multiple machine learning models are used to generate personalized recommendations for the various sections (for example, rows) of the Netflix homepage. Netflix's recommender system had a number of components prior to using deep learning. According to [45] breaking down the problem of recommendation into smaller subtasks allows us to combine multiple approaches and also makes the investigation and development of new or improved recommendation algorithms more modular [44].

In Figure 5.4, a Netflix homepage is shown. There are algorithms in place to help members find new videos, such as one that ranks previously viewed videos based on how likely it is that they will be re-watched. Each algorithm's output can be displayed on the homepage in the form of recommended videos in separate columns. Additional personalization options have been added to the Netflix service [44]. For example, a program determines which content rows should be displayed in a customized manner in order to construct the overall layout of the homepage [46]. In addition, the messages and notifications that are sent to members of the community are personalized. It also employs techniques based on user recommendations in its search engine [47].

5.5.1.1 Bag-of-Items

A bag-of-items assumption, which is a bag of videos in this context, is similar to the bag-of-words approach in NLP. Rather than focusing on the order in which a user watched videos, the model treats them as a collection. To represent training data in a bag-of-items setting, it is common to create a sparse user-by-item matrix where each non-empty entry represents the user's feedback on an item, both implicit (like plays or clicks) and explicit (like a rating or review, such as

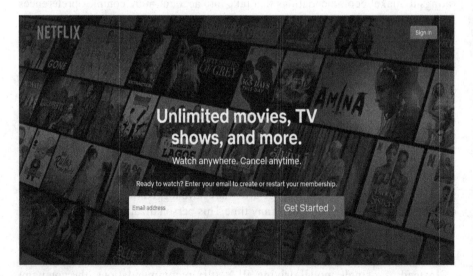

FIGURE 5.4   Netflix home page (Netflix.com).

a thumbs up or down). Aside from movie sequels and TV show episodes, there isn't much sequential information to be found in video engagements. If we want to model a user's long-term interests, we can use a bag-of-videos assumption that can provide a useful approximation to the real world in terms of temporal dynamics [44].

5.5.1.2 Sequential Models

It's convenient to have a bag-of-items view, but the sequential nature of a user's play history removes useful information. Many of the deep sequential models originally developed for NLP tasks, can be adapted to the recommendation task to accommodate this. When a user interacts with an item, the recommendation model aims to predict what they will interact with next, rather than what word they will use next. For session-based recommendations in e-commerce applications, sequenced methods are very effective: there is no clear user-identifier, so only the sequence of items visited so far is known. On these tasks, several models presented in [49–50] showed good performance.

While Netflix has used a variety of sequential models over the years, some of the more notable ones are long short term memory (LSTM) and gated recurrent unit (GRU) [48] and transformer architectures like BERT [49]. (Devlin et al. 2018). Despite the fact that transformers are not sequential models, their attention mechanism has the same effect. Additionally, to increasing recommendation accuracy, the attention mechanism introduces a novel and intriguing method of figuring out why a particular video is being suggested in the first place [44].

The use of machine learning techniques in Netflix recommender engines does not come without limitations. Notable limitations according to [44] include but are not limited to 1. mismatch in online and offline settings, and 2. overfitting on short-term proxy objectives that may be misaligned with longer-term objectives.

5.6 CONCLUSION

It is difficult for customers to make an informed decision in the face of so much information readily available to them online. There has been a lot of interest in recommender systems as a possible solution to the problem of information overload faced by knowledge workers and users. Many different recommendation systems have arisen as a result of this. The truth is that there is not a single approach that works best in every situation or for every person all the time. A number of studies have shown this to be the case. Recommender systems based on machine learning and artificial intelligence are a natural fit for the problems of traditional recommender systems. Fuzzy logic, evolutionary algorithms, and artificial neural networks were all discussed in this chapter in relation to how they can be used to model efficient recommender systems. AI/ML techniques in recommender system engineering have been shown in studies to be responsible for the current successes of recommender systems.

REFERENCES

[1] P. K. Singh, P. K. Dutta Pramanik, A. K. Dey, and P. Choudhury, "Recommender systems: An overview, research trends, and future directions," *International Journal of Business and Systems Research*, vol. 15, no. 1, 2021, doi: 10.1504/IJBSR.2021.111753.

[2] Mudita and D. Gupta, "A Comprehensive Study of Recommender Systems for the Internet of Things," in *Journal of Physics: Conference Series*, 2021. doi: 10.1088/1742-6596/1969/1/012045.

[3] A. Vineela, G. Lavanya Devi, N. Nelaturi, and G. Dasavatara Yadav, "A Comprehensive Study and Evaluation of Recommender Systems," in *Lecture Notes in Electrical Engineering*, 2021. doi: 10.1007/978-981-15-3828-5_5.

[4] I. Portugal, P. Alencar, and D. Cowan, "The use of machine learning algorithms in recommender systems: A systematic review," *Expert Systems with Applications20*, vol. 97. pp. 205–227, 2018. doi: 10.1016/j.eswa.2017.12.020.

[5] Q. Zhang, J. Lu, and Y. Jin, "Artificial intelligence in recommender systems," *Complex & Intelligent Systems*, vol. 7, no. 1, 2021, doi: 10.1007/s40747-020-00212-w.

[6] M. H. Mohamed, M. H. Khafagy, and M. H. Ibrahim, "Recommender Systems Challenges and Solutions Survey," in *Proceedings of 2019 International Conference on Innovative Trends in Computer Engineering, ITCE 2019*, 2019, pp. 149–155. doi: 10.1109/ITCE.2019.8646645.

[7] L. A. Zadeh, "Fuzzy logic - A personal perspective," *Fuzzy Sets Syst*, vol. 281, 2015, doi: 10.1016/j.fss.2015.05.009.

[8] L. A. Zadeh, "Fuzzy logic = computing with words," *IEEE Transactions on Fuzzy Systems*, vol. 4, no. 2, 1996, doi: 10.1109/91.493904.

[9] L. A. Zadeh, "Soft Computing and Fuzzy Logic," *IEEE Softw*, vol. 11, no. 6, 1994, doi: 10.1109/52.329401.

[10] A. Zenebe, L. Zhou, and A. F. Norcio, "User preferences discovery using fuzzy models," *Fuzzy Sets Syst*, vol. 161, no. 23, 2010, doi: 10.1016/j.fss.2010.06.006.

[11] R. R. Yager, "Fuzzy logic methods in recommender systems," *Fuzzy Sets Syst*, vol. 136, no. 2, 2003, doi: 10.1016/S0165-0114(02)00223-3.

[12] M. Mao, J. Lu, G. Zhang, and J. Zhang, "A fuzzy content matching-based e-Commerce recommendation approach," in *IEEE International Conference on Fuzzy Systems*, 2015, pp. 1–8. doi: 10.1109/FUZZ-IEEE.2015.7338036.

[13] D. Wu, G. Zhang, and J. Lu, "A fuzzy preference tree-based recommender system for personalized business-to-business e-services," *IEEE Transactions on Fuzzy Systems*, vol. 23, no. 1, 2015, doi: 10.1109/TFUZZ.2014.2315655.

[14] M. Nilashi, O. bin Ibrahim, and N. Ithnin, "Multi-criteria collaborative filtering with high accuracy using higher order singular value decomposition and Neuro-Fuzzy system," *Knowl Based Syst*, vol. 60, 2014, doi: 10.1016/j.knosys.2014.01.006.

[15] Z. Zhang, H. Lin, K. Liu, D. Wu, G. Zhang, and J. Lu, "A hybrid fuzzy-based personalized recommender system for telecom products/services," *Inf Sci (N Y)*, vol. 235, 2013, doi: 10.1016/j.ins.2013.01.025.

[16] R. Yera, J. Castro, and L. Martínez, "A fuzzy model for managing natural noise in recommender systems," *Applied Soft Computing Journal*, vol. 40, 2016, doi: 10.1016/j.asoc.2015.10.060.

[17] C. Cornelis, J. Lu, X. Guo, and G. Zhang, "One-and-only item recommendation with fuzzy logic techniques," *Inf Sci (N Y)*, vol. 177, no. 22, 2007, doi: 10.1016/j.ins.2007.07.001.

[18] L. H. Son and N. T. Thong, "Intuitionistic fuzzy recommender systems: An effective tool for medical diagnosis," *Knowl Based Syst*, vol. 74, 2015, doi: 10.1016/j.knosys.2014.11.012.

have attracted increasing attention due to their ability to support users in making better decisions and establishing users' faith in the system.

6.2 COGNITIVE RESPONSE TO EXPLAINABILITY IN AI

Relevant studies have empirically shown that the existence of an interpretable explanation develops user trust in the AI system and further argue that a user's trust is considerably influenced by the observed normative values that are used to judge algorithmic qualities (Sokol & Flach, 2020). Shin (2021) proposed a model of explainability AI that comprises the dual-step flow of users' cognitive processes of explainability in AI – namely, non-performance and performance attributes. In the model, explainability is posited as an antecedent of the tenets of fairness, transparency, and accountability. The model shows users' levels of algorithmic explainability and the effects of such attributes on attitudes and values. In this context, algorithmic attributes refer to the perceptions of algorithmic activity mediated by technology when encountered by readers, together with an interpretation of the complex ways algorithmic and user agencies emerge through interactivity.

The explainable AI model shows the facilitating and heuristic role of explainability in the user's dual-step flow of interaction with AI (Figure 6.1; Shin, 2021) – indicating that algorithmic information flows from AI to

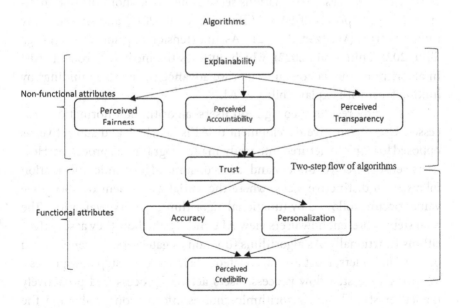

FIGURE 6.1   Dual-step flow of algorithmic explainability.

the user's evaluative framework of the procedural aspect of algorithms first, and from there to the performative aspect of algorithms: a peripheral-central route to evaluating the algorithm's features and deciding how and whether to continue to use AI. Algorithmic explainability is utilized to trigger a heuristic that users use to determine their values and attitudes toward AI services. The findings show that framing explanations can shape how people perceive and read news via AI-enabled chatbots. Holzinger et al. (2020) confirmed that explanatory cues can trigger positive and persuasive reactions by initiating and sustaining positive user heuristics. Trust mediates between the two steps by connecting explainability to the evaluation of quality held by users (Shin & Park, 2019). Users utilize explainability as an information cue method when discerning trust and quality in AI and determining their behavioral intentions. Explainability helps users understand non-functional algorithmic qualities, subsequently leading users to evaluate functional algorithmic performance and facilitate explanations (Rai, 2020). When users interact with AI, they make decisions as to what extent people accept AI-based reports as credible information sources, which is also related to concerns about how and to what extent readers should believe in algorithm-driven news (Shin, 2021). Such decisions are based on heuristic user judgments of normative values, including FAT, which are activated by explainability (Shin, 2021). In this process, explanations serve as heuristic shortcuts that influence user perceptions of FAccT (Thurman et al., 2019) and subsequently influence trust (Weitz et al., 2021). As an extension of prior research (e.g., Shin, 2021; Shin et al., 2022), which outlines the mediating role of trust in algorithmic media consumption, we advance the existing findings by linking trust with explainability in AI.

These study results shed light on users' algorithmic information processes. Users' cognitive development of AI is heuristic and subjective, as opposed to being structured into ready-made programmed processes. How users sense, perceive, understand, and consume algorithmic information follows two distinctive steps, where the initial assessment of normative values occurs, followed by functional algorithmic quality evaluation. The dual steps represent how users view AI ethically and how they assess algorithms functionally. As algorithms function as gatekeeping agents in lieu of opinion leaders, trust exerts a related role in the dual-step flow process. With the dual-step flow process, users actively process and proactively maneuver stimuli and algorithmic choices/information, evaluating the outcomes received in terms of normative values and trust. Trust in AI is cognitively formed in such a manner that processes are transparently explained and interpreted in human cognition. Readers are naturally skeptical of algorithmic processes, fearing that algorithms may not be as objective as discerning journalists and may instead evince race-based, sex-based, and other social biases of their human coders. Accordingly, users seek to know how algorithms work, how fake news and chatbots operate, and how to protect against disinformation. The two-step flow mechanism furthers the understanding of how algorithms influence decision-making by dividing the user's cognitive process into dual steps: (1) evaluating algorithmic values and (2) algorithmic attributes.

The dual-step flow in the model expands the literature on user experiences with AI – specifically user cognitive process research – by characterizing the role of trust and structuring the underlying relationships among its closely related measures (Dörr & Hollnbuchner, 2017). Understanding the extent to which users proactively manage sequences and processes in algorithmic decisions has key implications for theories and the operation of AI (Thurman et al., 2019). Findings on the role and/or process of explanations and the relationships among their associated measures not only confirm the theory's key argument – that cognitive decision is guided through a dual process – but also advance the theory by implicating this process in the two-step flow of communication. Users process algorithmic stimuli to evaluate whether they can understand and interpret the results of machine learning. When provided with interpretable explanations, readers feel more assured when evaluating algorithmic attributes. Explanatory cues help readers assess the tenets of FAccT and further establish trust, which in turn links and facilitates the processes between peripheral and central routes, enabling heuristic user activity, evaluations, and attitudes. Examining the interplay between procedural and performative processes in AI news clarifies that the two-step flow theory is relevant to AI research.

6.2.1 Mediating Effect of Explainability

The mediating effects of explanation offer key insights into user adoption of AI and algorithms (Rai, 2020). Research on the effects of explanation on trust and behaviors has consistently shown that explainability exerts a positive and important influence on adoption intention (Shin, 2021). Rosenfeld and Richardson (2019) confirmed that the presence and availability of explanations are a kind of user heuristic in AI. Arrieta et al. (2020) also found a relationship between explanation and trust. The presence of explanations – together with trust – is vital in improving media acceptance (Figure 6.2). Explanations have been turned out by numerous research studies to have significant effects on elucidating variations in user trust and adoption. Extending the role of trust, it can be hypothesized that the effects of FAccT on trust are facilitated by explainability – meaning that explanatory cues also affect user trust. Mediating effects can also be postulated between explainability and performance expectancy. Mediation analyses are designed to assess the mediating effects of explanation on trust and performance expectancy and to further test the mediating effects of explanation in FAccT and trust.

6.2.2 The Dual-Step Flow Model of AI Interaction

The two-step flow model of AI interaction (Shin, 2021) suggests that a user's trust has a far stronger influence than algorithmic processes on shaping user interaction with AI. This proposition is based on the dual-step flow of communication model (Lazarsfeld et al., 1944; Katz, 1957), which states that opinion leaders who are directly affected by mass media influence the perspectives of most people in society by passing on not only the media messages but also their own understandings thereof. While the theory has been widely accepted and applied to the mass media era, nowadays, its influence has diminished because there is a free flow of information such that anybody can have access to media content without the help of opinion leaders.

However, in the era of AI and algorithms, this traditional theory of communication has triggered new attention. Massive data are being used to generate recommendations or make certain decisions for users, reverting to the principle of a two-step flow of communication. In AI contexts where people and algorithms directly interact without intermediary agents or opinion leaders, users evaluate the quality of AI

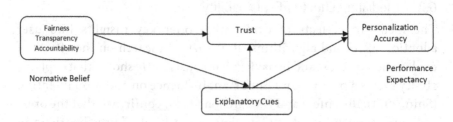

FIGURE 6.2 Mediation effects by explainability in AI.

messages by themselves and assess the trustworthiness of the messages or AI outcomes. The process of evaluating trust is greatly influenced by a user's own heuristics, which are also influenced by cues and other forms of stimuli that help users understand the processes of algorithms. In AI, ideas flow from algorithms to trusted users and from them to a wider group.

Based on the two-step interaction proposition, the term explainability illustrates the process intervening between the algorithm's direct recommendations and the user's reaction to those recommendations. Compared to the traditional dual-step flow of communication, where explainability can be seen in opinion leaders who are directly affected by mass media, explainability in AI reflects the internal logic of the algorithms, which can thus be analogous to opinion leaders in mass media. Users are influenced by the presence of explainability or other forms of attributes that are designed to help them understand. Trusted users tend to show loyalty toward AI and continue to use and change their attitudes and behaviors toward AI. Shin et al. (2022) conducted a series of experimental surveys on algorithmic users, and their findings supported that users using news recommendation services by AI-driven chatbots undergo a dual-step flow of interaction with AI. Within the interaction flow between AI and users, the trust mechanism plays a key role in creating, moving, and sustaining the interaction flow, and explainability triggers a trust evaluation. This dual-step flow of interaction implies some of the important underlying cues triggering and facilitating user interactions with AI, such as explainability, interpretability, fairness, and transparency, which all contribute to establishing user trust in AI. Further, the dual-step interaction refined the capability to predict how algorithmic recommendations influence user behavior and clarified why certain algorithmic interactions do not change users' attitudes and trust. The model stands in contrast to the technological determinism of algorithms, which assumes that algorithms determine user attitudes and behaviors and directly influence the way users accept algorithmic recommendations.

The two-step interaction can be a suitable argument for understanding the algorithmic influence on user attitudes and behaviors. Communication flows via algorithmic platforms find that today's AI media landscape concurrently facilitates a dual-step and more complicated multi-step flow mechanism of interaction (Figure 6.3).

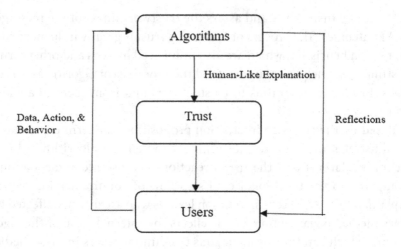

FIGURE 6.3   Two-step model of human-algorithm interaction.

6.3 STANDARDS FOR EXPLAINABLE AI

Advanced and sophisticated AI algorithms nowadays are black boxes. While they can perform well, their inner workings are mysterious and unexplainable (Reviglio & Agosti, 2022). AI that uses deep neural networks is so complicated that even its programmers cannot understand the inner workings. Due to concerns about the black box nature of algorithms, there is an increasing call for action on explainability and transparency. Explainability is considered a key part of trustworthy AI since it provides an understanding of how a model behaves and where its use is appropriate. The pervasiveness of bias and susceptibilities in AI systems means that trust is unwarranted without an adequate understanding of how a system works.

The National Institute of Standards and Technology has defined four doctrines of explainable AI: explanation, significance, precision, and limits of knowledge. The principle of explanation obligates the AI system to offer an explanation in the form of justification or evidence for each outcome.

The principle of significance defines that the subject of the explanation should be comprehensible and customized to the user, both at the individual and organizational levels. In terms of precision, a specific explanation should be provided to show how the system generates the final result. Because the application of precision depends on the situation and the users, different measures of explanation precision will be presented for different types of users and communities. For the fourth principle, AI operates only under the conditions for which it was intended or when the system gets suitable confidence in its outcomes. The principle of knowledge limits obliges the system to address any cases for which it was not designed.

These four principles indicate that AI-based services must have the required explainability and transparency to generate trust in their functioning and confidence in the results of the system.

The IEEE Society for Computational Intelligence and the Committee on Standards issued the Standard for Explainable AI in 2021. This standard specifies obligatory and discretionary requirements and limits that should be satisfied for an algorithm, machine learning method, application, or system to be explainable.

As there is currently no standard that provides a single high-level methodology for classifying AI products as partially or fully explicable, there is a great need for it, as researchers, scientists, and developers of AI systems are limited by their specific services, users, and conflicting interests. The problem becomes more critical when interoperability comes into play. A single standard would allow us to optimize requirements and quality, increase productivity, improve the quality of the final service, and fulfill the needs of users.

6.4 A RIGHT TO EXPLANATION

Algorithms perform in manners that are often mysterious, even to their own designers. Algorithms generate unknowns – an analytics of user data that are understood in certain terms, such as the patterns and relations from the data set, but incomprehensible to others, such as the context that the data embodies. For example, when a credit card firm considers a certain transaction or series of purchasing as possible fraud and places a transaction hold on the card, the firm would not explain to the customers clearly what was dubious or suspicious, although they know that there is something strange. They would not and could not reveal because they do not know themselves what it is. The right to explanation is increasing as algorithms become more sophisticated and opaquer and less understandable (Shin, 2021).

The EU GDPR mandates that users have a "right to explanation" of decisions made by algorithmic systems. Article 22 of the GDPR ("Automated Individual Decision-Making, Including Profiling") states, "The data subject shall have the right not to be subject to a decision based solely on automated processing, including profiling, which produces legal effects concerning him or her or similarly significantly affects him or her." In addition to the general notion about the right to explanation of specific algorithmic decisions, the GDPR states that data subjects should obtain meaningful information (Articles 13–15) about the method used and the importance and expected results of automated decision-making systems based on their "right to be informed." Despite the narrow application and loosely worded mandate, the GDPR serves as a basis for requesting general explanations of almost all algorithmic systems. Nowadays, under the policies of AI, algorithms, and machine learning, users have a legal right to obtain an explanation of a result from an algorithm, meaning data subjects have the right to be given meaningful information about how their data are used and how the data lead to certain decisions, in the same way that a person who applies for a job and is rejected may ask for an explanation. There have been arguments that a social right to explanation is a fundamental basis for an information society, particularly as the organizations of that society will use AI, algorithms, and machine learning. Without this right of explanation, the public will have no way to challenge the decisions made by algorithms.

The principle of the GDPR has been occurring in different forms across the world. For example, the CCPA in the US gives state residents the right to opt out of the sale of personal data, to have collected data disclosed, to be informed of data reuse, and to equal services and prices. In the private sector in the US, creditors are obliged to provide applicants who are rejected credit with reasons for the decision under the Equal Credit Opportunity Act (Regulation B of the Code of Federal Regulations, Title 12, Chapter X, Part 1002). Creditors comply with this directive by giving a list of reasons, consisting of a numbering reason identifier and a related explanation, finding the key factors influencing a credit score. Besides the Equal Credit Act, many US States mandate the Right to Know Act, which is designed to ensure public access to the public records of governmental agencies in the states. For example, the Pennsylvania Sunshine Act enacts methods by which public meetings are conducted. California's Right to Know Act of 2013 (Assembly Bill 1291) authorizes citizens to gain complete disclosure upon request of personal data held by organizations about the citizens.

In France, the 2016 Digital Republic Act established the nation's administrative code to initiate a new regulation to explain the decisions made by public agencies about citizens. Although the regulation is limited to governmental administrations, it goes beyond the GDPR's right in that the regulation holistically requires broad explanations surrounding decision processes. It defines where a decision is made on the basis of an algorithmic process; the process and its main characteristics should be provided to the citizen upon request. Except for national defense and security reasons, this regulation is widely applied. Specific explanations should be included: (1) the data analyzed and its origin; (2) the mode and the degree of contribution of the algorithmic processing to the outcome; (3) the operations carried out by the process; and (4) the used parameters and their weigh importance applied to the case of the citizen concerned.

One of the rising issues is how to give an explanation to a user and how an explainable system can be understood by users. In complicated deep learning models, it is difficult to explain clearly why the software made a prediction in a particular case. This lack of traceability has made organizations hesitate to use AI in critical fields, such as judiciary systems, healthcare, and finance. Criticism and opposition have created a challenge for explainable AI. Many industry proponents have voiced that the right to explanation is destructive and threatens innovation as the rule restricts developers to a standard that is infeasible and unnecessary in many cases. It has also been argued that the regulation overlaps other regulations, has an unbalanced focus on process over results, and favors human decisions over algorithm decisions. From a technical perspective, the current algorithms utilized in AI are mostly difficult to explain. For instance, the results of a machine learning model depend on multiple layers of analyses connected in a complicated manner, and no single input or data may be a contributing factor. Just like the human decision process, where people often depend on a gut feeling, it is difficult to dissect algorithmic decision-making into pieces and articulate it. Given the challenges and issues raised, a group of researchers proposed explainability from a wider question raised by the social right to an explanation instead of an overly limited focus on scientific or product specifications (Shin et al., 2022). Such a social right to explanation is notable because the right to an explanation has been an individual right, yet the scope of a general right to an explanation is a matter of illusive debate. A social right to explanation can be a key foundation for an AI society, particularly as the institutions of that society will need to use AI, algorithms, and machine learning. Algorithmic decision-making systems that include explainability would be more reliable and transparent. Without this right, the public will be left without much alternative but to question the decisions of AI systems. One of the rising questions is how to communicate an explanation to the public. Should it be through a document, a visualized log diagram, video, or some other medium, and how can an explainable system scope the explanation in a reasonable way?

6.5 APPLICATION IN EXPLAINABLE AI USE CASES

6.5.1 Explanatory Journalism

Explanatory journalism is the practice of reporting to present news stories in an understandable way by allowing greater context than would be presented in conventional journalism (Mann, 2016). A recent example of explanatory news is Vox, which provides analytic journalism by explaining complex issues or events as understandable storytelling.

Traditionally, the focus of journalism has been on reporting news promptly and accurately. Thus, attention has been paid to producing the speedy coverage of news objectively rather than explaining the news. As AI is applied to journalism, it is becoming increasingly important to explain the news to digital readership. Traditional journalists have the sole right to decide what news is provided, create tones, and write articles in ways that impact reader acceptance. In traditional journalism, there is little room for readers to demand explanations, which creates both ethical and practical concerns, particularly with the application of AI to journalism. As journalism embraces algorithms, explainability becomes key in algorithmic journalism; wherein editorial decisions are based on user behavioral data. Explainability is necessary because the core goal in algorithmic journalism is to establish credibility by rationalizing how and why journalistic practices work. This explainability is so critical that it has even been legislated in many cases. Explainability in journalism enables readers to comprehend how and why algorithms generate particular news. Therefore, newsrooms have increasingly started to explain their editorial decisions to counter decreasing levels of trust.

As users' rights to know have been increasing, news audiences may wonder about the justifications for why and how curated news is processed and generated (Ferrario et al., 2020). Explainability is key in creating faith, credibility, and trust between AIs and users, especially when it comes to countering clickbait, fake news, and disinformation (Rai, 2020). Explainability is particularly important in computational/digital journalism, wherein credibility and trust are critical issues. Explainability in algorithmic journalism affords users trust and confidence that AI systems perform well, facilitating an understanding of why a system operates in a certain manner and safeguarding against bias and prejudice. Explainability also allows users who interact with algorithms to make accurate and relevant suggestions in algorithmic journalism. Therefore, explainability is the key to the safe, ethical, fair, and trustworthy use of algorithms in journalism and is a key enabler of its deployment in the journalism domain. Journalists will soon face a compelling need to justify and editorialize their news articles.

6.5.2 News Recommendation Systems

AI algorithms have made it possible for developers to create news recommendation systems (Shin, 2020a). Recommender systems, which are enabled by algorithms, help users access ever-growing sets of services and data available on the internet (Moller et al., 2018). The user chooses news to view or articles to subscribe to, and the system then proposes items that may potentially interest the user based on his/her preferences or previous viewing history. Thus, users are given recommendations as a result of products that they have already viewed or rated.

Reading personalized news online has become quite common, as the web offers unlimited access to news items from many sources (Shin, 2020b). Because the volume of news articles can be overwhelming to users, building a news recommendation system to help users curate the articles that are most appealing to them is a key task for online news services (Thurman et al., 2019). With the advancement of algorithm technologies, news recommendations have been widely adopted and will be diffused further in society (Figure 6.4). These systems shape profiles of users' news preferences based on their behavior on the internet. The systems aim to

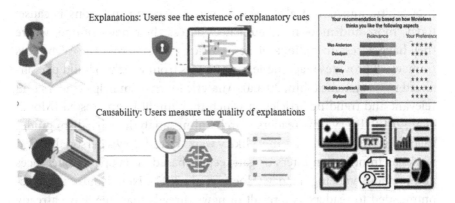

FIGURE 6.4   Example of explainable AI in journalism.

identify news that best fits users' preferences and are often considered a primary source of news articles for users (Konstan & Riedl, 2012).

6.5.3 Analytic Platforms

Across industry, user analytics are widely adopted and used in platform businesses. Amazon employs different kinds of explanations in its AI-based systems. Some recommendations are marked with a sentence stating how it was discovered, such as "Customers who purchased this item also purchased . . . ." Amazon has also made it possible for users to decide how much a purchase should influence their recommendations or whether it should be filtered out altogether. This explanation improves the auditability of the system and can be useful for filtering out items purchased as gifts. Customers can also understand which of their previous purchases or assessments affected a recommendation.

Facebook provides an explanatory cue that advises why users are viewing a particular post, as well as suggestions on why the posts are organized in the way they are. Users are also able to read itemized information about why a commercial was shown to them, including a timeline for when and how the advertising creator has interacted with their personal data. By sharing such information, Facebook attempts to boost transparency in its services and allow users to control their own news feed, i.e., improve the auditability of the service. The Korean online platform Naver uses text word clouds of feature – opinion pairs to explain why certain news is recommended. Research found that the explanatory cue users who saw the text explanation were significantly higher than those who received a generic explanation (e.g., "People also considered . . .") or no explanation at all (Shin et al., 2022).

Analytic journalism also uses AI-driven recommendations because, today, more and more news audiences access their news online, where they have access to billions of news articles from across the Web. This framework provides machine-learning bases from which to build personalized news systems (Shin, 2020a). Analytic journalism helps readers find relevant and trending articles in which they might be interested (Moller et al., 2018). Every time readers search online, analytic journalism guides them toward the news they most likely want to read. News personalization works as information filters or gatekeepers based on people's preferences and previous viewing history (Hoeve et al., 2017). News sources are recommended to readers as a result of news threads that they have already viewed or interacted with, such as rating stories or tagging posts. Major news agencies, such as the BBC, CNN, *The Washington Post*, and *The New York Times*, are adopting recommendation algorithms to improve user experiences with their newsfeeds. The functions that recommend personalized and relevant news articles to readers in a journalistic context are different from those of other recommendation domains. One lingering issue relates to a central question regarding the extent to which journalistic gatekeeping processes will be replaced by algorithms (Moller et al., 2018). There has been mounting criticism over the fairness and transparency of AI services, mandated by the truthfulness of journalism, in terms of the goals, services, and methods of algorithms utilized to search for, process, and generate recommendations (Shin et al., 2020). Rising concerns over transparency and fairness have resulted in a demand for more reasonable and understandable explanations in AIs (Dörr & Hollnbuchner, 2017). These issues can substantially challenge algorithm and AJ services by producing a set of negative and even disruptive effects in AI systems (Rai, 2020). Explainable recommendations, which provide explanations about why an item is recommended, have drawn increasing attention in AI and journalism due to their ability to support users in making better decisions and to establish users' faith in the systems and credibility of algorithmic journalism.

6.6 BRIDGING THE GAP BETWEEN EXPLAINABILITY AND HUMAN COGNITION

In the early stages of AI development, people accepted it even if they did not understand what the model predicted in a certain manner, as long as it produced accurate outputs. Explaining how algorithms function was not an important consideration. However, the increased application of AI has increased the complexity and scalability of the systems, and accordingly, the need for visibility, understandability, and explainability of the AI-based system has drastically increased. The question, then, is whether such requirements are meaningful for the users of algorithmic services. The focus is now turning from technical explainability to providing human-interpretable models. Explainability is an inherently human-centered property, and the industry has begun to accept human-centered approaches. The quality of explanation or explainability lies in the perception and acceptance of the human experiencing the explanation. In addition, this shift of focus is related to the current lack of explainability in AI. Most of the explanations provided by algorithms are not useful for users to understand the inner workings of algorithms. Earlier attempts of explainability were to make a machine learning model entirely visible to its variables and data, but industry quickly realized that making things visible does not ensure that the user at the receiving end can make sense of all the information. What makes an explanation effective depends on the users' existing schema and their intentions for receiving the explanation, among other human factors. Thus, designing explainable models necessitates human-centered principles that highlight the technical development of users' understandability needs and evaluate explainability success through human empowerment, engagement, and informativeness. The goal of the interpretable AI model is to allow users to draw a rough outline of how the model performs because it is impractical to know the complicated logics of a black box algorithm model. This is why the human cognitive process has garnered great attention, and a research community of human-centered explainability has risen, which brings in cognitive perspectives. A group of researchers has proposed the idea of interpretability as an alternative concept of explainability (Holzinger et al., 2019). Shin et al. (2022) defined interpretability as the extent to which the internal logics of AI can be explained in understandable human cognition and language. Interpretability is associated with how accurately algorithms can associate a cause with an effect, whereas explainability is about the ability of the parameters to justify the results (Shin, 2021). Along with explainability, interpretability is crucial in establishing trust and understanding between an AI agent and its user, particularly in the case of undesirable consequences and unanticipated blunders (Arrieta, 2020). For example, in the context of algorithmic journalism, interpretability allows editorial staff, reporters, and newsrooms to better understand why certain news algorithms are generated and amend them as needed.

Renijith et al. (2020) researched the interpretation of AI systems from a human perspective and identified causality, transferability, trust, information, and fair decision-making as key factors contributing to the interpretation of AI. Holzinger et al. (2019) summarized the taxonomy of explainability in relation to cognate variables, such as interpretability, explicitness, transparency, and faithfulness. Rosenfeld and Richardson (2019) proposed a review of the system's transparency in an algorithm framework, highlighting the role of transparency in flexible and efficient human–robot interactions. While transparency is becoming increasingly critical, the abstract quality nature of transparency, along with explainability, should be better understood in relation to the mechanisms that can promote it. Anjomshoae et al. (2019) argued that an important first step is to identify user requirements in terms of interpretable AI behavior. Ongoing works on explainable AI have commonly argued that what we really need is understandable AI instead of explainable AI. Their common concerns can be persuasive, given that explainable AI has not been a key solution for all problems that have occurred. The technical rationales should be driven by users' interpretability requirements, for which theories of human cognition and behaviors can provide conceptual bases to inspire new computational and design frameworks for interpretable AI. The ideal of understanding how AI makes decisions is great, but specific explainable methods or approaches have never been sufficient (Holzinger et al., 2019). We should consider how to warrant full trust in algorithm systems' performance by addressing how to deliver interpretable AI. In this light, Shin et al. (2022) proposed key aspects of interpretable AI: transparency, causability, the ability to question, and ease of understanding. Transparency is the basis of understanding. Every decision made by machine learning models should be open to non-technical users. These users should be able to check a database based on important factors to assess judgments both institutionally and individually. They should also be able to perform a counterfactual analysis of individual decisions, altering particular factors to see if the outcomes are as predicted. Interpretable AI mixes developers' technical knowledge with user experience experts' design usability understanding, as well as interface designers' human-centric approach. Users can engage in the decision-making process in AI-based institutions if the algorithms are open. The inclusion of social scientists and cognitive researchers in the design and development of AI services is key to the interpretable AI process, illustrating the importance of the human approach to interpretation. A seamless fusion of humans and algorithms is necessary because human intervention should be superseded by algorithmic modeling to allow humans to assess the quality of decision paths and decrease false positives. Interpretability should involve some level of human-in-the-loop intervention. The conditions of visibility, understandability, and interpretability are not wholly technical matters; they are representations and outputs of human cognitive processes (Shin et al., 2022).

6.7 BEYOND EXPLAINABLE AI

Impressive progress in machine learning and the rise in algorithmic power have led to the development of AI systems that can be used to diagnose disease and make financial investments, job applications, or autonomous vehicles. However, the potential of AI systems is restricted by their inability to provide explanations to users. While AI cannot be developed to be faultless, firms need to be able to explain to users how AI is being used so that related bias risk can be alleviated, trust can be built, and meaningful benefits can be gained.

Despite it being truly appealing, there have been problems presenting explainability in AI. The most significant difficulty is that interpretable, trustworthy, and explainable AI has not yet been realized in practice despite efforts by the explainable AI field. Nowadays, a considerable discrepancy exists between the goal of explainability to become a norm that reaches across fields and works for diverse stakeholders and how it is being used in practice. The struggle comes mostly from the wide definitions of what explainability is supposed to realize, which creates confusion about the unequal prioritization of various stakeholder goals. Recent research has reported that the excessive implementation of explainable AI may constrain the functionality of AI (Bruijn et al., 2022). Current explainable AI systems have focused mostly on designing AI systems that are explainable to AI specialists rather than regular users, and their results on user cognitions of the algorithms have been rather inconsistent (Shin, 2022). Gunning et al. (2019) reported that technical dimensions are generally prioritized over other aspects, with explainability mainly falling short of the needs of users, stakeholder groups, and communities. For technical reasons, explanations at the human level are very hard to obtain for users as well as for developers to achieve. Arrieta (2020) suggested using fundamentally interpretable algorithm coding rather than adding post hoc explanations, where an additional model is added to justify the initial results. Post hoc explanations complicate a decision tree, and it is often vague how accurately a post hoc explanation can represent the computations of a completely detached model. Despite its limitations and challenges, explainable AI will continue to evolve, and there is great opportunity and value in this explainable system. The adoption of explainability will continue to become more critical for all AI sectors.

In developing AI, the common goal of making AI understandable to users should be shared across different stakeholder groups, from designers, developers, programmers, and users. Without the common goal of explainability among various stakeholders, AI is more likely to serve the interests of only the powerful groups in societies. AI firms should thus clarify how they are using explainable methods, to what end, and why, and make full transparent explanations. The regulatory bodies developing explainability standards and regulations should take into account the current limitations of explainable AI in practice and look for diverse expertise about how to better align governance and incentives with a full picture of explainable AI goals. Only with the active engagement of broad stakeholders, from computer science, social sciences, industry, and community groups can we achieve the vision of trustworthy, understandable, and controllable AI in practice.

REFERENCES

Anjomshoae, S., Najjar, A., Calvaresi, D., & Främling, K. (2019). Explainable agents and robots. *Proceedings of the 18th International Conference on Autonomous Agents and Multi-Agent Systems*, 1078–1088. Retrieved from www.ifaamas.org

Arrieta, A. B. (2020). Explainable artificial intelligence: Concepts, taxonomies, opportunities and challenges toward responsible AI. *Information Fusion*, *58*, 82–115.

Bruijn, H., Warnier, M., & Janssen, M. (2022). The perils and pitfalls of explainable AI: Strategies for explaining algorithmic decision-making. *Government Information Quarterly, 39*(2), 101666. https://doi.org/10.1016/j. giq.2021.101666

Bucher, T. (2018). *If . . . then: Algorithmic power and politics*. New York: Oxford University Press.

Chazette, L., & Schneider, K. (2020). Explainability as a non-functional requirement: Challenges and recommendations. *Requirements Engineering, 25*(4), 1–20. doi:10.1007/s00766–020-00333-1

Combs, K., Fendley, M., & Bihl, T. (2020). A preliminary look at heuristic analysis for assessing artificial intelligence explainability. *WSEAS Transactions on Computer Research, 8*, 61–72. doi:10.37394/232018.2020.8.9

Dörr, K. N., & Hollnbuchner, K. (2017). Ethical challenges of algorithmic journalism. *Digital Journalism, 5*(4), 404–419. doi:10.1080/21670811.2016.1167612

Ehsan, U., & Riedl, M. O. (2019). On design and evaluation of human-centered explainable AI systems. *Glasgow'19, Scotland*, ACM.

Ferrario, A., Loi, M., & Viganò, E. (2020). In AI we trust incrementally. *Philosophy & Technology*. https://doi.org/10.1007/s13347-019-00378-3

Gunning, D., Stefik, M., Choi, J., Miller, T., Stumpf, S., & Yang, G. (2019). XAI: Explainable artificial intelligence. *Science Robotics, 4*(37), 7120. doi:10.1126/ scirobotics.aay7120

Hoeve, M., Heruer, M., Odijik, D., Schuth, A., Spitters, M., & Rijke, M. (2017). Do news consumers want explanations for personalized news rankings? *FATREC 2017*, August 31, Como, Italy. https://doi.org/10.18122/B24D7N

Holzinger, A. (2016). Interactive machine learning for health informatics: When do we need the human-in-the-loop? *Brain Informatics, 3*(2), 119–131. doi:10.1007/s40708-016-0042-6

Holzinger, A., Carrington, A., & Müller, H. (2020). Measuring the quality of explanations: The System Causability Scale (SCS). *Künstl Intell, 34*, 193–198. https://doi.org/10.1007/s13218-020-00636-z

Holzinger, A., Langs, G., Denk, H., Zatloukal, K., & Mueller, H. (2019). Causability and explainability of artificial intelligence in medicine. *Data Mining and Knowledge Discovery, 9*(4). doi:10.1002/widm.1312

Katz, E. (1957). The two-step flow of communication: An up-to-date report on an hypothesis. *Public Opinion Quarterly, 21*(1), 61–78. doi:10.1086/266687. JSTOR266687

Konstan, J. A., & Riedl, J. (2012). Recommender systems. *User Modeling and User-Adapted Interaction, 22*(2), 101–123.

Lazarsfeld, P., Berelson, B., & Gaudet, H. (1944). *The people's choice.* New York: Columbia University Press.

Mann, T. E. (February 2016). Explanatory journalism: A tool in the war against polarization and dysfunction. *Brookings Institution.*

Moller, J., Trilling, D., Helberger, N., & van Es, B. (2018). Do not blame it on the algorithm: An empirical assessment of multiple recommender systems and their impact on content diversity. *Information, Communication & Society, 21*(7), 959–977. doi:10.1080/1369118X.2018.1444076

Rai, A. (2020). Explainable AI: From black box to glass box. *Journal of the Academy of Marketing Science, 48*, 137–141. https://doi.org/10.1007/s11747-019-00710-5

Renijith, S., Sreekumar, A., & Jathavedan, M. (2020). An extensive study on the evolution of context-aware personalized travel recommender systems. *Information Processing & Management, 57*(1), 102078. https://doi.org/10.1016/j.ipm.2019.102078

Reviglio, U., & Agosti, C. (2022). Thinking outside the black-box: The case for algorithmic sovereignty in social media. *Social Media + Society.* doi:10.1177/2056305120915613

Rosenfeld, A., & Richardson, A. (2019). Explainability in human-agent systems. *Autonomous Agents and Multi-Agent Systems, 33*(6), 673–705. https://doi.org/10.1007/s10458-019-09408-y

Shin, D. (2020a). How do users interact with algorithm recommender systems? *Computers in Human Behavior, 109*, 1–10. https://doi.org/10.1016/j.chb.2020.106344

Shin, D. (2020b). User perceptions of algorithmic decisions in the personalized AI system: Perceptual evaluation of fairness, accountability, transparency, and explainability. *Journal of Broadcasting & Electronic Media, 64*(4), 541–565. https://doi.org/10.1080/08838151.2020.1843357

Shin, D. (2021). Why does explainability matter in news analytic systems? Proposing explainable analytic journalism. *Journalism Studies, 22*(8), 1047–1065. doi:10.1080/1461670X.2021.1916984

Shin, D. (2022). The perception of humanness in conversational journalism: An algorithmic information-processing perspective. *New Media & Society.* doi:10.1177/1461444821993801

Shin, D., & Park, Y. (2019). Role of fairness, accountability, and transparency in algorithmic affordance. *Computers in Human Behavior, 98*, 277–284. doi:10.1016/j.chb.2019.04.019

Shin, D., Zaid, B., Biocca, F., & Rasul, A. (2022). In platforms we trust? Unlocking the black-box of news algorithms through interpretable AI. *Journal of Broadcasting and Electronic Media.* https://doi.org/10.1080/08838151.2022.2057984

Sokol, K., & Flach, P. (2020). Explainability fact sheets: A framework for systematic assessment of explainable approaches. *Conference on Fairness, Accountability, and Transparency,* Barcelona, Spain. https://doi.org/10.1145/3351095.3372870

Thurman, N., Moeller, J., Helberger, N., & Trilling, D. (2019). My friends, editors, algorithms, and I. Examining audience attitudes to news selection. *Digital Journalism, 7*(4), 447–469. https://doi.org/10.1080/21670811.2018.1493936

Vallverdú, J. (2020). Approximate and situated causality in deep learning. *Philosophies, 5*(2), 1–12. doi:10.3390/philosophies5010002

Weitz, K., Schiller, D., Schlagowski, R. et al. (2021). "Let me explain!": Exploring the potential of virtual agents in explainable AI interaction design. *Journal on Multimodal User Interfaces, 15,* 87–98. https://doi.org/10.1007/s12193-020-00332-0

CHAPTER 7

Algorithmic Journalism: Current Trends and Future Developments

ALGORITHMS STRUCTURE HOW WE see and think about the world around us and have been used to make decisions about most areas of human life. Moreover, advances in artificial intelligence (AI) have enabled process automation, which has led to new information services, such as content generation, information curation, news recommendation, and content optimization. Based on these services, algorithms have transformed journalism in terms of news production, newsroom structure, and overall journalistic activities. Journalists around the world are figuring out how to make use of algorithms to improve user experience and journalism services. Using Naver's AI-based recommendation system as a case study, this chapter discusses the methods and services of algorithmic journalism, showing how an algorithm functions in news services, how the algorithm is used, processed, and understood in different journalistic contexts and via different tools and approaches, and how it is communicated to users. For algorithmic journalism to be sustainable, algorithmic designers should understand journalistic values and integrate them into algorithm construction. Algorithmic journalism involves serious ethical considerations regarding fairness, transparency, accountability, and explainability.

DOI: 10.1201/b23083-8

7.1 INTRODUCTION

Recently, news stories have come to be generated automatically by algorithms rather than manually by human journalists (Diakopoulos, 2019). This trend, commonly referred to as algorithmic journalism, has garnered sensational popularity due to radical technological advances in the domain of journalism (Lewis et al., 2019). Algorithmic journalism can take many forms and can provide many different services, such as robot journalism, conversational journalism, and chatbot journalism (Figure 7.1; Zamith & Haim, 2020). Recently, scholars have started considering algorithmic journalism beyond automatic content generation, focusing, instead, on issues of human and journalistic values. For

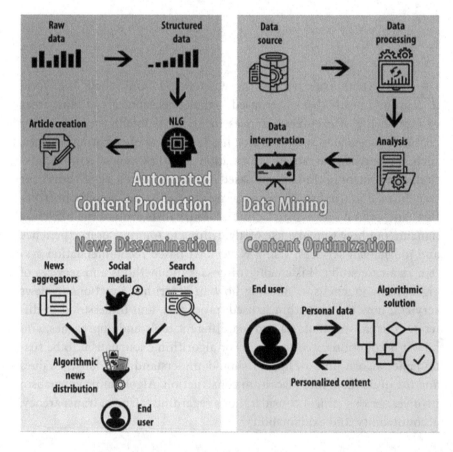

FIGURE 7.1   Areas of application for algorithmic journalism (see Kotenidis & Veglis, 2021).

example, Shin (2020) defined algorithmic journalism as the integration of process, data, and knowledge with algorithms to maintain the fairness and accountability functions of journalism. Conceptualizations of this sort consider a large variety of innovative technological services that can be applied in journalism while recognizing the essential role of human journalists in such processes. Similar innovations, such as computer-assisted reporting, have been employed by journalists in their practice, including the use of databases and conducting data analysis to provide context and depth to news stories. Unlike these conventional innovations, algorithmic journalism is revolutionary because it transforms journalism values, norms, and practices by automating search functions, classifications, and data processing (Shin, 2022). Algorithms shape all stages of the journalistic workflow and process, such as collecting data, organizing information, framing news, and reporting stories. However, the algorithmization of the news production process has generated heated debates over the control of algorithms in relation to journalistic values (Wölker & Powell, 2021). With radical advances in artificial intelligence (AI) technologies, algorithmic journalism is increasingly capable of performing jobs that were once the sole responsibility of human journalists, and the repercussions of this shift have triggered a discussion on whether future newsroom work will be totally programmed as part of technological progress (Caswell & Dörr, 2018; Deuze & Witschge, 2018). These AI-induced shifts take various forms, from the streamlining of processes, such as collecting relevant data for news stories, to performing more complicated tasks, such as writing news articles from scratch with packaged software like WordSmith (a writing robot that produces 2 billion articles per year), to the extent that every phase of the news production process can be done by algorithms without human supervision.

7.1.1 Algorithmic Filtering and Gatekeeping

How Platforms Use Algorithms to Decide What to Show to Users

Most of the current online platforms embed algorithms to rank, filter, suggest, and control content, thus favoring some information at the expense of other information. Algorithms use curation techniques to organize, select, analyze, and process a body of information for recommendations. Curating techniques involve filtering, selecting, prioritizing, classifying, and bracketing data, which are the components of algorithmic mediating processes (Shin, 2021). Platforms such as Netflix, Google, Amazon, Naver, Twitter, Facebook, and other news aggregators act as algorithmic gatekeepers and curators. Machine learning algorithms embedded in recommender systems collect and evaluate large data sets to make personalized recommendations to target users. These processes can be described as algorithmic intermediation between people and information that creates shortcuts and streamlines interaction. Algorithms are the new intermediaries performing the function of gatekeeping and filtering in the context of algorithmic journalism. As this kind of gatekeeping power can exert tremendous influence over society, researchers and policymakers have advocated algorithmic audits, a systematic approach to evaluating algorithms to see how they work and whether they are functioning according to their stated goals or generating biased results (Smets et al., 2021). These days, almost all research institutes dealing with AI ethics have called for algorithmic audits. For example, the US Congress's Algorithmic Accountability Act of 2022 requires "Automated Decision System Impact Assessments" in all decisions made by algorithms and machine learning. The landmark bill requires new accountability and transparency measures from almost all algorithmic decision-making systems. Similarly, the EU High-Level Expert Group on AI has highlighted the need for "Ethics Guidelines for Trustworthy AI." However, how algorithmic audits can be conducted remains an open question and is an area of ongoing debate. In practice, it is very difficult to access the algorithms used. Recent regulatory proposals have been too hypothetical to be put into practice without additional procedural details. Practical limitations exist as to what algorithm firms can be required to provide. For example, somebody whose reputation has been damaged by fake news and misinformation should be given a reasonable explanation and should be entitled to request corrections to the articles. At the same time, news organizations should not be obliged to compromise intellectual property rights by fully disclosing algorithmic details.

Industry guidelines on algorithmic audits focus on the technical conditions of transparency, fairness, and accountability, lacking connection to the broader social context. While the need for algorithmic audits has been widely recognized, providing a mechanism for such audits is a complex challenge. Given their profit-seeking nature, companies are unwilling to share their algorithms and, therefore, shield them from public scrutiny. Algorithmic firms should be required by law to audit and make visible the systems applied in their operations. Challenges in actionable algorithmic audits that could lead to greater accountability persist due to algorithms' proprietary features, which are considered private intellectual properties

(Diakopoulos & Koliska, 2016). Even if full transparency is offered, the complexity of technical algorithms remains a barrier to understanding their impacts. Moreover, algorithms may vary and may respond to input or output in ways that cannot be analyzed. In real AI systems, even for a single online application, there is no single code that could be audited; instead, there is a web of countless interconnected paths and aggregated parameters (Shin et al., 2022).

7.1.2 News Algorithms: Algorithmed Public Spheres

News recommendation systems, or news algorithms, recommend the most pertinent news article to readers based on readers' personal preferences and interests (Karimi et al., 2018). The primary goal of personalized services is not to serve more users but to serve them better, thus establishing stronger relationships and attracting loyal customers. Algorithmic software creates journalistic content that can be valuable, especially in light of information overload, as too much information of little relevance causes confusion (Beam & Kosicki, 2014). Users choose the news to read or content to buy, and the system then suggests articles that may possibly interest users based on various data sources, such as previous purchases and data history. Algorithmic software identifies readers with similar news interests based on their behavior, such as retweeting articles on Twitter or "liking" items on Facebook. Therefore, users are offered recommendations based on their interactions with algorithms, to which they provide data and which analyze the data for personalized curation. Algorithmic recommendations are widely used in news algorithms, which has made reading news online even more common. Accurately predicting user needs and effectively curating and personalizing content are important issues for media platforms (Cotter, 2019). Personalizing media by matching content and news to users' preferences can enhance reader engagement and revenue streams for platform providers. The development of machine learning technologies has further increased the diffusion of algorithms in the media. News algorithms have been widely adopted and will soon become a mainstream trend. The ways in which news algorithms identify possible news have become increasingly accurate and scientific, as algorithms can correctly profile users' news preferences based on a multitude of data sources. Consequently, algorithms are nowadays seen as the main source of news for readers (Shin, 2021).

Despite AI's enormous popularity and obvious benefits, such as improved efficiency through automation and higher user engagement via scientific filtering, there are concerns regarding privacy, fairness, bias, and transparency (Diakopoulos & Koliska, 2016). Algorithms are not a perfect remedy for the presence of bias in journalism and society; in reality, if algorithms are flawed, they can amplify bias and exacerbate discrimination and inequality. Examples of such problems include fake news, mis/disinformation, and so on. There are many instances when algorithmic journalism becomes self-reinforcing and vulnerable to manipulation. Concerns over manipulation, in particular, have received significant criticism from policymakers and the public (Ananny & Crawford, 2018). There is a need for transparency in algorithm services, and firms need to make visible the strategies, structures, and procedures underlying the algorithms used to search for, process, and deliver information. For example, in 2021, Facebook announced increased transparency in its algorithmically curated news feed to improve user experience and balance content quality with user rights. Global platform providers have started to realize the importance of transparency because it is closely related to user engagement and ethical issues. In fact, without transparency, algorithm-based services can create undesired and even hazardous problems in AI systems (Diakopoulos, 2019).

As algorithms control platforms, various questions have emerged: Who controls the algorithms behind the news that people read? How does personalization impact editorial values? What rights do users have? Algorithms have contributed to an increasingly polarized society and have increased people's ability to read only what interests them, thus intensifying selection bias and limiting people's ability to encounter different perspectives. Nowadays, the public wants to be better informed about the process behind news quality, as users believe they have the right to access diverse recommendations (Helberger et al., 2018).

In other words, the public is requesting that something be done about the black box nature of algorithms – that is, the problem of not knowing the processes that algorithms rely on to make decisions (Burrell, 2016). Simply put, people do not know how and why algorithms produce recommendations and whether the results are correct. The black box issue is a problem in news recommendations and entails fundamental ethical and legal questions: (1) Are recommendation results fair and accurate? (2) How can we ensure the transparency of algorithmic processes?

These questions are related to ethical issues as well as managerial and operational principles of algorithm journalism. The aforementioned ethical problems mean that fairness, accountability, and transparency (FAccT) issues are a key research area in AI scholarship. In algorithmic services, FAccT issues have been a thorny subject (Diakopoulos, 2019). The FAccT principle constitutes not only operational values but also ethical issues to be resolved in AI design. The principle is based on the realization that AI is not fair and just. Therefore, the FAccT principle is about preventing bias and the unintended consequences of algorithmization. When content is recommended algorithmically, it should be easy to trace who is legally or politically responsible for any potential harm to make sure that recommendation decisions do not have an unjust or discriminatory impact on users. Without this principle, users will have increasing difficulties accepting algorithmic services.

7.1.3 Growing Need for Algorithmic Fairness and Transparency

Algorithmic journalism involves ethical and legal questions without clear answers. Transparency and objectivity are critical values in journalism and media and have also been deemed essential in news recommender systems (Diakopoulos, 2019). These concepts are frequently considered in the design and development of news recommender systems and algorithmic journalism (Dörr & Hollnbuchner, 2017). Recently, algorithmic journalism has started to apply the FAccT principle in the design of news automation (Shin, 2021). As the FAccT principle has been successfully used to address the black box problem in machine learning, the principle can also be employed to tackle the ethical issues surrounding news algorithms in the journalism domain. Given that news algorithms are normally invisible – in fact, algorithms are often referred to as "black box" processes because they are not visible to users, and their code is generally not public – most people who rely on algorithms every day are ignorant of how they work and why they can pose risks. When news recommender systems provide selected news, it remains unclear how the filtering processes work and whether the recommended news actually corresponds to user preferences (Just & Latzer, 2020). As these issues are important in journalistic practice, efforts are underway to resolve and prevent damaging or unjust news-selection processes. The FAccT principle is rapidly growing in importance as journalism attempts to untangle these complex problems using diverse perspectives on how best to proceed technologically and socially as well as journalism-wise. These challenges need to be seen in relation to the core values of journalism, such as justice, accuracy, fairness, and media responsibility. Fairness in traditional journalism and fairness in algorithms may vary according to content and context. Algorithmic journalists face enormous challenges regarding the automated processing of big data and the continuous updating and incorporation of machine learning to facilitate smart news selection that responds to ever-changing user needs and expectations, such as receiving truthful, high-quality, and plural information that is, at the same time, personalized. In addition, misinformation and fake news are often generated and circulated by algorithms, and algorithmic journalists are held responsible for such mishaps.

In the face of various challenges, algorithmic journalists have strived to tackle the related ethical concerns, which have arisen very suddenly. However, it should be noted that FAccT issues can be best addressed from a user perspective (Shin et al., 2022). Users want trustworthy systems that tell them how the data are examined and thus how the recommendations are produced. When processes are transparent, users can revise inputs to improve news recommendations. Moreover, users of news recommender systems can then understand the principles and processes of news filtering and gatekeeping, and news recommendation providers can ensure that the results are legitimate and genuine. When fair and transparent services are provided, users are likelier to consider the news as credible and of high quality. Highly transparent algorithms can provide users with assurance, and accurate news fosters users' sense of trust. Finally, high visibility and transparency increase search performance and user satisfaction with the system.

7.2 CASE STUDY OF NAVER'S ALGORITHMIC NEWS

The case of Naver's algorithmic news service offers an important and cautionary paradigm for algorithmic journalism regarding the dynamic relations between algorithmic curation of news and the conceptualization of related issues.

7.2.1 Algorithmic Journalism in South Korea

Users in South Korea predominantly consume news and information through media platforms, such as Naver and Kakao (Kwak et al., 2021). Naver, the fourth largest search platform in the world with 30 million monthly active users, is considered Korea's Google (Dwyer & Hutchinson, 2019). Most Koreans access the majority of their news on the Naver app or website (Shin, 2020). People rely on Naver, Daum, and Kakao for news services, as these platforms offer highly sophisticated news personalization services. Naver leads online news consumption with 68% of the user share, and Kakao is in second place with 32% based on a combination of news, chat, and email (Kwak et al., 2021). The high reliance on Naver and Kakao has made these platforms the dominant news providers in the country (Kim & Moon, 2021). As these platforms use AI technology in their news-aggregating services, concerns about human bias and the technical reliability of the news-ranking mechanism have been raised (Bhadani et al., 2022). The platforms have been criticized for manipulating or "maneuvering" the news article rankings in favor of certain political parties (Bhadani et al., 2022). In light of this criticism, in August 2018, Kakao introduced the first AI-driven news recommender system: Real-Time User Behavior Interactive Content Recommender System (RUBRICS). Following this trend, in June 2019, Naver developed its AI Recommender System (AiRS), an in-house algorithm that provides user-based personalized news. RUBRICS uses a combination of machine learning and tailored multi-armed bandit algorithms, while AiRS uses a recurrent neural network algorithm (RNNA), a deep learning mechanism that traces users' behavioral data and the temporal sequences in which users view news content.

These algorithms have greatly increased news clicks and traffic by personalizing news articles according to user preferences and interests. AI-based news recommender systems can assuage increasing public criticism regarding political bias. Platforms need customer data in deciding the choice of particular news topics for coverage and the timing of an article's release. With AI prevalence in the newsroom, platform providers seem to have almost legitimate control over creating and presenting news to the public (Thurman et al., 2019). Gatekeeping has thus become a key issue in AI-driven news in South Korea. News algorithms play the role of digital intermediaries by curating data and conducting gatekeeping activities. For Koreans, Naver and Daum are content platforms using which they can search for information, read news based on their preferences, and receive suggestions and guidance regarding items to search for (Dwyer & Hutchinson, 2019). These platforms serve as social hubs where users can browse information, with lots of quality content that may then nudge users to search for related news on topics of interest. More specifically, Naver is the primary information source for the majority of Koreans; moreover, the platform has a very strong relationship with Korean publishers and news agencies. However, active AI gatekeeping has prompted concerns that news algorithms may manipulate what people see online, potentially negatively affecting public opinion by presenting biased views or recommending a narrow range of topics (e.g., due to filter bubbles and echo chambers). The public has raised the issue that full information is not provided to readers or that readers are exposed to only partial or extreme information (Kwak et al., 2021). In fact, one study argued that news algorithms in Korea produce partial information blindness (Shin, 2021).

7.2.2 Naver News Algorithms: AI-Driven News Recommendations

One of the most prominent algorithms in the country is the news ranking and recommendation system used by Naver, the dominant platform provider in the country. Naver has become the major destination for news, replacing the traditional journalism industry. In 2019, Naver developed AiRS, a personalized news recommender system, which has been used by the platform's news services since then. AiRS is equipped with algorithms that recommend news items based on the following two criteria: (1) a quality model that automatically assesses article quality using a set of standards and user feedback and (2) a collective filter that analyzes user groups based on preferences. Naver's AI system analyzes users' intentions and tastes and helps users find relevant information through search results for various topics.

The service simultaneously provides readers with popular, non-personalized headline recommendations (current topics, breaking news, and hot issues) and, in a specific part of the web page, displays a personalized list of news snippets. Unlike other AI news systems, AiRS uses a combination of contextual information and collaborative filtering approaches, which helps to minimize inaccurate results and unrelated recommendations while considering position and layout bias to provide more accurate and relevant recommendations. AiRS is used in almost all of Naver's services, including video, discussion, news, and cartoon services. Naver offers a list of the most-viewed news items in different sections – politics, society, and economy – based on the number of clicks or comments. With AiRS, the company has been developing its news recommendation algorithm to consider qualitative criteria, such as popularity, relevance, and other users' viewing patterns rather than simply tracking numbers. The AI system uses deep learning, collaborative filtering, and reinforcement learning to avoid algorithmic errors and improve the accuracy of the recommendations. AiRS also integrates large-scale data refinement and serving methods enabled by the yet another resource negotiator technique (data flow software), which is capable of processing a maximum of 12,000 transactions per second. In 2022, Naver updated its AI with Hyperclova, a new data processing AI system.

7.2.3 How AiRS Works

The recommendation system mainly identifies users' long-term preferences using algorithmic methods, such as tensor factorization of the "(user x news x context)" tensor, which is systematically recorded and processed. This tensor tracks the history of a user-news log, showing which user viewed which news at what time, for how long, and in what contexts. For individual users, the recommendation system also identifies short-term intentions by establishing, in an online and incremental manner, a reader profile based on a low-dimensional representation of the sequence of clicked and unclicked news items. Both short-term and long-term user preferences are recorded as user relevance scores, which can be converted into tailored attractiveness measures between a user and a possible news article to be suggested. Other criteria, such as context (location and/or time of access), the characteristics of the pictures related to the news article, and readership trends, are then calculated with user-specific multi-temporal weight measures using an "orchestrator" learning-to-rank algorithm. This orchestrator algorithm employs previous search and/or recommendation records to best forecast the news items with the highest probability of being clicked on. The AiRS system tracks users to capture periodic searching and/or browsing patterns, such as checking baseball results every weekend, and browsing habits, such as reading about an event related to an article that was accessed previously and that made the user want to explore more.

Different perspectives are considered in the recommendation list by including clusters of news items that recommend related news or other perspectives reported by different news outlets. This provides users with access to a wide spectrum of key news and events that may interest them, with the possibility of exploring other opinions associated with the various facts and viewpoints. This enables users to delve into specific news (inter-cluster diversity) and browse diverse views and arguments regarding a specific news item (intra-cluster diversity). In 2021, based on AiRS, Naver introduced Hyperclova, an ultra-large AI model comparable to Elon Musk's OpenAI (Figure 7.2).

7.2.4 Concerns Regarding News Algorithms

Since its inception, the news ranking service has achieved sensational popularity (Bhadani et al., 2022). The dependency on online news services in South Korea is much higher (86%) than the average dependency in Western countries (54%). Current digital platforms in the country tend to

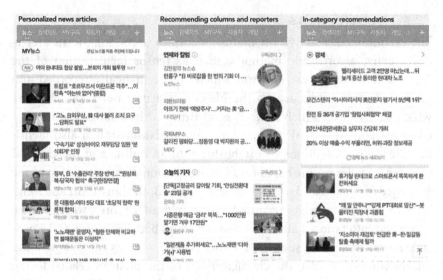

FIGURE 7.2 The AiRS news recommendation system.

be monopolistic, with the majority of users relying on only two platforms. Moreover, users have experienced algorithmic manipulation. Scholars have argued that Korean readers rely heavily on algorithm-driven platforms for news access (Shin, 2020). More than half of Korean users access news primarily through platforms (e.g., Naver or Daum), and about 10% of users consume news through social media services, such as KakaoTalk (Newman et al., 2020). A survey done by the Korea Press Foundation indicated that 80% of respondents stated that they consumed news via platforms, which rerouted readers to the original news articles through links. Due to the high acceptance of news algorithms in the country, the majority of Korean users (62%) consider platforms to be journalism: they see platforms as alternative media, communal content hubs, and conduits between users and news.

In addition to platforms' increasing influence in shaping the Korean news ecology, concerns have been raised regarding the functioning of news-ranking algorithms, particularly their trending search chart feature. For example, Naver implemented an "AI editor" based on personalization algorithms (Kim & Moon, 2021); the AI editor filters and recommends news stories based on users' news search records. This feature provides users with curated packages of news stories that are personalized based on users' interests and preferences according to an examination of news consumption patterns. The AiRS algorithm is designed to predict users' preferences by considering other users with similar news browsing patterns. The RNNA is embedded in the AiRS system to identify viewers' behavioral data and the orders in which users read news articles. One problem related to the embedded functions is "position bias," whereby the higher-ranked item is viewed more often, even when several news items are evenly relevant. This problem is coupled with other biases, such as systematic bias, whereby users click on the topmost suggestion since it is generated by the system, or the layout prejudice, whereby items with suggestive preview information, such as clickbait news, are likelier to be clicked on.

Naver's AI design has been criticized for creating or at least enhancing a clickbait media environment. The clickbait design makes it difficult to distinguish an accurate relevance signal from noisy click information. Furthermore, Naver has been criticized for amplifying societal biases and creating filter bubbles or ideological isolation in that their algorithms selectively frame what information users see. These bias problems have increased public doubts about news algorithms. Against the backdrop of the ever-growing influence of platforms' news algorithms on users, the level of public trust in the news services offered by platforms has been the lowest among the different news providers, largely due to the lack of fairness, transparency, and, recently, explainability in their platforms' operations (Shin, 2021). The transparency of algorithmic curation has deteriorated due to human editors who have attempted to manipulate and rig the rankings of the most popular news articles, sowing distrust of the rankings among the public. Naver, the prominent platform provider in the country, has received the most criticism regarding algorithmic transparency and fairness. Since 2013, Naver has been accused of maneuvering its own real-time news search ranking service. To respond to the suspicions of manipulation, Naver has attempted to keep the real-time search word ranking transparent by voluntarily requesting that the Korean Internet Self-Governance Organization supervise and evaluate the service on a regular basis. Following the public backlash over the opinion rigging concerns, Naver established a series of rules, such as restricting the number of likes from a single account and regulating online comments. Amid continued criticism from lawmakers and the public, in 2018, Naver finally increased the visibility of the real-time service by placing it on a subpage that was reachable after at least five page clicks from the front page. Despite these efforts, allegations regarding the manipulation of rankings to sway public opinion remain to this day. Naver has also faced charges that it manipulates its search algorithms in favor of preferred advertisers affiliated with the company – in fact, in 2021, the Korean antitrust authority fined Naver $24 million for misleading users by using its market dominance to illegally block competitors (Kim & Moon, 2021).

Although Naver has publicly rejected accusations of news-ranking manipulation, it is clear that rankings should depend solely on neutral algorithmic filtering, and the debates on potential manipulation have continued to this time day. Critics argue that algorithmic firms should reveal the algorithms behind the news selection process by allowing a third-party audit of the code – in other words, it is not enough to simply say that the program is working properly. In its defense, Naver has claimed that the news algorithm system curating the news rankings works automatically, visibly, and impartially. However, Naver has not made public the key conditions behind its algorithmic ranking or the extent to which human judgment was involved in the process. Korean news platforms, including Daum and Kakao, have not revealed what data sources were fed into what kinds of algorithmic mechanisms to prioritize some news pieces while filtering out others. Amid the long controversy over fairness and transparency, Naver decided to end its online real-time search chart service in March 2021.

7.2.5 Algorithmic Transparency and Fairness

Although South Korea has boasted technological prowess in terms of AI and algorithm development, significant progress has not been made in clarifying the specific rules and regulatory standards that would facilitate the transparency and accountability of news recommendation algorithms (Shin, 2019). Although Naver has the highest market share of news consumers, the company has been at the center of public criticism regarding the algorithmic transparency of its news-ranking system. Naver has been frequently accused of manipulating its own real-time news search ranking service. To respond to the criticism of tweaking, Naver has attempted to keep the real-time search service transparent by allowing a third party (namely, the Korean Internet Self-Governance Organization) to audit the service. Although the real-time search service has been one of their most popular services, Naver finally agreed to reveal the inner mechanism of the real-time service by making it accessible on a subpage that was reachable after a minimum of five-page clicks from the homepage. Although this open policy required considerable effort, allegations of online opinion rigging and ranking manipulation did not abate. Naver has never accepted the allegations of news rigging. The company has claimed that the news rankings and suggestions involve purely algorithmic operations and depend entirely on algorithmic curation. Nonetheless, public concern regarding possible manipulation has been increasing due to the company's refusal to reveal the key criteria and core algorithmic processes behind the rankings, as well as the extent to which human intervention was involved in the selection process. Just like global platform providers, such as Google and Facebook, Korean platform providers have not released information on algorithmic operations (types of data sources, the process of machine learning modeling for prioritizing news selections, and the gatekeeping mechanism). In fact, almost all platform providers in Korea have never been open about the principles, criteria, and quality of curated news, and little effort has been invested in checking the veracity and reliability of such criteria.

In 2021, the country's major platforms agreed to define and accept algorithmic accountability and established ethical guidelines for algorithmic journalism. These initiatives can be seen as progressive because companies have started to realize the importance of algorithmic accountability and transparency. However, further work is needed to establish detailed procedures and measures for evaluating how algorithms operate, assessing the legitimacy of news algorithms and human interventions, and determining the extent to which providers are accountable for negative outcomes, such as fake news and mis/disinformation.

7.2.6 Wider Impacts

The findings indicate that algorithmic journalism has produced significant changes in South Korean journalism and society. Naver's algorithmic journalism service highlights the weaknesses of news algorithms, thus problematizing the prospect of algorithms serving as perfect alternatives to classical journalism. Although algorithms function as efficient and valuable search tools for users, they often amplify prejudices and exacerbate distrust in journalism. Naver's AI system revealed a series of issues to do with FAccT in algorithmic journalism. Based on the case study of the South Korean experience, it is possible to trace connections with the larger media ecology.

7.2.7 Fairness, Accountability, and Transparency (FAccT) in Algorithmic Journalism

Algorithms are becoming more powerful, sophisticated, and ubiquitous. However, rapid advances in AI technology have produced not only unprecedented opportunities but also new concerns among users, professionals, and government officials. Against the fast-changing technological backdrop, several social and ethical issues remain contentious (Kemper & Kolkman, 2019). How can fair algorithms be designed and implemented? How can we develop algorithms that are fairer and more transparent? In much of the current debate on credibility in algorithmic journalism, FAccT issues are frequently espoused without specific knowledge of how these issues are related to trust.

Algorithmic journalism has become widely popular in South Korea. Innovations in AI technology have resulted in highly sophisticated algorithmic journalism, ushering journalism and the platform industry into a novel era with new players and power structures. Advancements in machine learning provide unprecedented opportunities for breakthroughs in content curation, media services, and communications (Guzman & Lewis, 2020). Although there are many ways in which algorithms and AI are used in the media sphere, algorithmic journalism has faced ethical and societal concerns because it relies on data and technical assumptions and is subject to biases and unfairness. Some algorithmic journalism services promote problematic practices, such as attention-grabbing news and clickbaiting, that are ultimately harmful to users, such as filter bubbles, echo chambers, and clickbaiting (Shin et al., 2022). Once a story is promoted by news algorithms, a sudden increase occurs in its viewership. AI-based news systems have been proven to have a self-reinforcing nature and are easily vulnerable to manipulation (Beer, 2017). When these kinds of problems occur, the issue is who is responsible for the consequences of such negative effects. Given that algorithms cannot be held legally accountable, human accountability needs to be embedded in every stage of the process. Platform providers, such as Naver, Daum, and Kakao, hold great power in informing and possibly rigging public opinion, and the platforms, as owners of the means of production, thus, have an obligation to prevent the dissemination and promotion of damaging information by the algorithms that they have deployed. Clear measures should be prepared to stop intentionally misleading content. In addition, concerns have increased regarding the transparency of algorithm services, and platforms need to be transparent and open about the rules, structures, and procedures behind the algorithms used to search for, process, and provide information (Dwyer & Hutchinson, 2019). A lack of transparency and fairness can significantly challenge algorithm-based services by generating a series of undesired and even critical problems in AI systems (Diakopoulos, 2019).

The algorithmic journalism community needs to be prepared to handle FAccT issues, as related problems will continue to increase and impact the market. Discriminatory and biased algorithms may pose serious risks for journalism. How can we develop algorithmic journalism in accordance with the FAccT principle? How can this principle be reflected in the interfaces, how will people perceive it, and how will society respond to it? It is evident that automation can have serious negative effects in the form of fake news, mis/disinformation, and deep fakes. As algorithms translate data into narrative news stories automatically and in real-time, there is a risk of amplifying misinformation and fake news. Moreover, the algorithm cannot explain its reporting output: why it wrote what it did or how it got there. Such lack of clarity is the main source of algorithmic bias, which is increasingly amplified in negative data feedback loops. The ways in which data are collected, curated, and stored have a significant impact on the news algorithms that are trained using the data, and each step must agree with the FAccT principle. It is important to consider how legitimate and unbiased data can be collected, how the data are used for analytics, and whether the recommending processes are fair and impartial. Making algorithmic journalism accountable to readers is critical as algorithmic journalism develops.

Fairness in journalism means that reporting should be accurate and truthful and should not plant stories that cause readers to draw predetermined conclusions (Diakopoulos, 2016). In reality, journalists never succeed in being impartial and fair in presenting all sides of a story. This limitation is also present in algorithmic journalism, insofar as data collection, analysis, and presentation are supervised and/or intervened upon by human reporters. Although algorithms perform data-related processes automatically, it is human journalists who decide the scope of the data, the means for analyzing the data, the selection criteria, and the way in which data results are presented (Smets et al., 2021). Thus, algorithmic fairness in algorithmic journalism should involve identifying the potential causes of unfairness and discrimination at every stage of the algorithmic process (Shin & Park, 2019). At every step of the algorithmic process – for example, when extracting information from data sources, editing, and publishing – fact-checking and bias-removal practices need to be implemented. Algorithmic journalists should ensure not only the deployment of fairer algorithms but also design human procedures to decrease biases in the data. Such procedures may combine algorithms and humans in decision-making. It is important to ensure that human biases do not influence news algorithms.

Contemporary social media and online search platform algorithms are deliberately opaque. The concept of transparency in the context of algorithmic journalism refers to the principle that the processes and factors behind AI decisions should be visible to users and the public (Bishop, 2019). Recently, scholars have argued that transparency should be accompanied by easily understandable explanations of AI decision-making processes (Shin, 2020). In other words, algorithmic transparency is underpinned by the concepts of fairness, accountability, visibility, explainability, and interpretability. When transparency exists, people can see fairness; with transparency in place, users consider platforms more accountable; when transparency is ensured, people can check the inner algorithm workings; and when explainable interpretations are provided, people consider transparency to be high. Unsurprisingly, transparency has been described as the most important factor among all algorithmic attributes (Moller et al., 2018). However, algorithmic transparency in algorithmic journalism poses difficult challenges for platforms, professionals, and policymakers. Algorithmic transparency requires platform providers to open up about how algorithmic filtering supports recommendations by providing information on algorithmic software, algorithmic codes, and human-supervised decisions. As an independent private field, journalism can scarcely afford to publicly share valuable intellectual property. Also, even when AI-related processes are fully open, whether regular users can understand the complex algorithmic processes remains questionable (Shin, 2020). Despite conceptual ambiguity and operational inconsistency, algorithmic transparency has been actively pursued in journalism and the public sector. In the European Union, the data protection laws (2019) include the "right to explanation" regarding decisions made by AI. This right to explanation, similar to the right to reply in journalism (the right to respond to public criticism in the same venue where the criticism was published), empowers users to request an understandable explanation of the decisions and results produced by algorithms. Future journalism should implement this right to an explanation as a common and legal practice.

Algorithmic accountability involves identifying and assigning responsibility for harm when algorithmic journalism produces negative and discriminatory outcomes (Moller et al., 2018). There has been mounting public pressure to hold the providers of automated decision systems responsible for the results generated by their algorithms. The Data Accountability and Transparency Act of 2020 claimed to protect users' privacy rights and required high legal accountability from algorithm providers. In this accountability obligation, algorithmic actors (e.g., platform providers, journalists, and content providers) have an obligation to explain and justify their practices, designs, and/or decisions related to algorithmic reporting and the subsequent effects of such reporting. Although algorithmic accountability remains to be operationalized and enforced, journalism industries, as well as AI industries overall, need to answer pointed questions about accountability since recent trends in fake news and misinformation have turned into serious risks. These significant challenges lend urgency to the development of transparent, fair, accountable, and, therefore, trustworthy and acceptable algorithms.

7.2.8 User Role in the Formation of Algorithms: The Changing Concept of Users

From news aggregators to Google news, Netflix recommendations, and Instagram feeds, the way in which readers see information depends on the gatekeeping practices of platform algorithms. Although humans seem to be passive recipients, scholars have recognized users' cognitive processes and active roles in forming news algorithms. The rise of algorithms has produced a paradigm shift in how media companies see and classify their audiences. Whereas traditional media assumed that users belonged to particular social categories, algorithmic media see active, performative, participative, and collaborative partners based on behavioral data. With the rise of machine learning technologies, users' role has changed from that of passive recipients of automated processes through algorithms or media to active architects of algorithms who initiate, regulate, modify, and control such algorithms depending on the frames and contexts of their everyday lives (Bishop, 2019). Users want to see what they prefer to see, they want to view what they prefer to view, and they want to be empowered by the algorithmic process (Kotenidis & Veglis, 2021).

By referring to subconscious cognitive processes, many researchers have argued that users are the sources of algorithms and the creators of news recommendations. What users see through algorithms, as far as their cognition is concerned, is a cognitively constructed reality that emulates the form of accumulated experience that has been shaped by a priori mental constructs. As Just and Latzer (2020) argued, algorithmic filtering has become a shared social reality shaping daily lives and realities and affecting people's perceptions of the world. Humans and algorithms are co-evolving and co-constructing reality together as they mutually influence each other.

While algorithms reproduce user preferences, the negative effects of such a reflection have societal ramifications at various levels. At the micro level, creating user-centered algorithms is a matter of creating algorithms that are more transparent and responsible. The notion of accuracy is not a matter of reflecting what users want but of suggesting socially and politically correct information to users. This point suggests that user perceptions of transparency and accuracy are not purely objective responses to media content. Perceived transparency and accuracy in algorithmic media are, indeed, in the eye of the beholder. There are various dimensions using which we can measure how "transparent and accurate" a recommendation is. Such issues are socially constructed and cognitively reconstructed within users' cognitive dimensions. Rather than such issues being uniformly or collectively given to users, users create their versions of transparency and accuracy based on their levels of trust and other personal intrinsic factors. These lines of research stem from human-computer interactions. Algorithmic journalism would benefit from incorporating conceptual frameworks such as human-AI interaction and human-computer interaction (Lewis et al., 2019). Future research on algorithmic journalism would benefit from a broader methodological and theoretical scope to better capture the changing nature of the interaction between journalists and technology in most contemporary journalism.

7.3 CONCLUSIONS: SHOW ME THE ALGORITHM

The case of Naver's AI system shows the innovative services of algorithmic technologies that are applied to journalism: automated content generation, data processing, news dissemination, and personalized news delivery. Although these technologies have already revolutionized the ways in which journalism newsrooms work, there are numerous challenges that need to be addressed, not all of which are algorithmic or technological. Algorithmic journalism requires news organizations to figure out what readers want, what is technically viable, and what is legally allowed. The Naver case shows that, in addition to the apparent algorithmic shortcomings, various managerial, editorial, and ethical concerns have emerged, which implies that the ecology of algorithmic processes is both promising and challenging. As fairness and ethical concerns reach new levels of intensity, it is critical for algorithms to be as fair, transparent, and accountable as possible to continue their operations in tandem with desirable social values as well as conventional journalistic practices.

The overall influence of giant news platforms, such as Naver, is far more substantial than simply offering a convenient news digest for readers. The platform has the power to design how news is represented, shown, and then consumed by Korean users. Using AI-enabled power, Naver's dominant status in the news market has been strengthened by the unrivaled scale and personalized algorithms of its news service. At the same time, the case of the Naver AI system shows the importance of the FAccT principle, which has become a key concern regarding the uses and effects of recommendation algorithms. There is mounting pressure to increase algorithmic transparency, fairness, and accountability. However, how and to what extent this can be done remains an open question, as these values involve multi-dimensional, highly complex collections of regulations and processes as well as people's ability to understand. Even if users are invited to peek into algorithms, it is likely that many of them will not understand the functioning of algorithms, let alone the social implications of transparency. Nevertheless, efforts to ensure that algorithms are as fair and transparent as possible must continue.

7.3.1 Suggestions for News Algorithms

The importance of algorithms and automation in journalism and media has been growing. More and more news organizations are implementing AI technology for a variety of services. The discussion in this chapter can help journalists use AI technology in gathering, composing, and distributing news.

The first practical suggestion is that pertinent industries should address user experience related to algorithms and news recommendations. Subjective perceptions and psychological effects are critical in how users perceive and think about the services and implications of news recommendations and how they use and engage with algorithm-generated news. The main goal of news curation is to help people reach news that is interesting and intriguing to read. Understanding how users search, find, and consume news online allows algorithm providers and algorithm designers to perform their work more effectively. There are numerous challenges to offering recommended results in the journalism context (Kitchin, 2016). FAccT issues have been prominent in algorithmic journalism and in relation to recommendation systems overall. User experience is critical in making recommendations more accurate. The discussion of this chapter provides AI designers with guidelines on how to integrate fairness and transparency issues with other factors.

The second practical suggestion is related to trust and credibility. Trust in algorithmic processes is critical, and the industry should strive to earn user trust. FAccT issues are closely related to users' trust, which is key in promoting user satisfaction and trust. When users are assured of solutions to troubling issues, their trust levels increase, and they become more willing to allow more of their data to be collected and processed. To increase the trust between users and algorithms, more transparent processes are needed, and more data would enable algorithms to produce accurate results, tailored and individualized according to users' preferences and personal histories. Trust is a key factor in positive feedback loops between users and algorithm systems (Kemper & Kolkman, 2019). Moreover, scholars have argued that trust plays a key role in AI adoption, particularly in complicated algorithmic systems (Alexander et al., 2018). In a highly dynamic and hyper-connected environment, characterized by information overload and uncertainty, trust is a key factor in decision-making. When users trust certain services, they tend to believe that the services are underpinned by a transparent and fair process (Cramer et al., 2008). Therefore, a trust algorithm is a set of rules that enhances trust (Alexander et al., 2018). Higher satisfaction implies greater trust, and users are likelier to continue to use and adopt AI systems. Increasing user trust and control may assure users that their individual data will be used for legitimate and transparent processes, thereby generating perceived trust in the service and its providers and eventually leading to a heightened level of satisfaction.

The final suggestion concerns changing journalistic roles and values. To what extent are journalistic values and functions modified by algorithmic journalism? No matter how algorithms evolve and advance, algorithmic journalism will require human journalists. As the line between human journalists and algorithms is blurred, it is urgent to reappraise the roles played by AI and human actors. It should let human journalists do human tasks and algorithms do AI things – that is, humans and algorithms should work together harmoniously. Algorithmic journalism should incorporate the best of human judgment along with editorial values, with safeguards built in to prevent AI systems from turning into clickbait mills. Algorithms have no capacity to make moral decisions and thus need human journalists to make contextual *in situ* determinations and judgment calls. For example, does a specific story have the right context? Reporters need to verify the data to ensure their accuracy before feeding the data into the writing robot. Human journalists also need to give algorithm-generated news a final evaluation. Therefore, human journalists still play a key role in automated tasks, as algorithmic journalism relies on human contextualization and insights. In fact, algorithmic journalism should be a man-AI marriage in which the two collaborate instead of competing. Although the algorithmic process can produce personalized news within minutes, readers still desire the special human touch, such as emotions and empathy, that algorithms cannot offer. Journalists can normalize algorithms by adapting them to existing journalistic practices and values and, perhaps, can even redefine normative ideological paradigms by creating new values. The ultimate goal of news algorithms and recommendation systems should become human-centered services. User-centered algorithms and trust-based feedback loops will be critical in developing such human-centered systems. However, we must better understand how the interaction between algorithms, journalism, and news users shapes the flow of news.

REFERENCES

Alexander, V., Blinder, C., & Zak, P. (2018). Why trust an algorithm? *Computers in Human Behavior, 89*, 279–288. doi:10.1016/j.chb.2018.07.026

Ananny, M., & Crawford, K. (2018). Seeing without knowing. *New Media and Society, 20*(3), 973–989. doi:10.1177/1461444816676645

Beam, M. A., & Kosicki, G. M. (2014). Personalized news portals. *Journalism & Mass Communication Quarterly, 91*(1), 59–77. doi:10.1177/1077699013514411

Beer, D. (2017). The social power of algorithms. *Information, Communication & Society, 20*(1), 1–13. doi:10.1080/1369118X.2016.1216147

Bhadani, S., Yamaya, S., Flammini, A., et al. (2022). Political audience diversity and news reliability in algorithmic ranking. *Nature Human Behavior, 6*, 495–505. https://doi.org/10.1038/s41562-021-01276-5

Bishop, S. (2019). Managing visibility on YouTube through algorithmic gossip. *New Media & Society, 21*(11–12), 2589–2606. doi:10.1177/1461444819854731

Burrell, J. (2016). How the machine thinks: Understanding opacity in machine learning algorithms. *Big Data & Society, 3*(1), 1–12. doi:10.1177/2053951715622512

Caswell, D., & Dörr, K. (2018). Automated journalism 2.0: Event-driven narratives. *Journalism Practice, 12*(4), 477–496. doi:10.1080/17512786.2017.1320773

Cotter, K. (2019). Playing the visibility game: How digital influencers and algorithms negotiate influence on Instagram. *New Media & Society, 21*(4), 895–913. doi:10.1177/1461444818815684

Cramer, H., Evers, V., Ramlal, S., van Someren, M., Rutledge, L., Stash, N., Aroyo, L., & Wielinga, J. (2008). The effects of transparency on trust in and acceptance of a content-based art recommender. *User Model User-Adapt Interact, 18*(5), 455–496.

Deuze, M., & Witschge, T. (2018). Beyond journalism: Theorizing the transformation of journalism. *Journalism, 19*, 165–181.

Diakopoulos, N. (2019). *Automating the news: How algorithms are rewriting the media*. Boston, MA: Harvard University Press.

Diakopoulos, N., & Koliska, M. (2016). Algorithmic transparency in the news media. *Digital Journalism, 5*(7), 809–828. doi:10.1080/21670811.2016.1208053

Dörr, K., & Hollnbuchner, K. (2017). Ethical challenges of algorithmic journalism. *Digital Journalism, 5*(4), 404–419. doi:10.1080/21670811.2016.1167612

Dwyer, T., & Hutchinson, J. (2019). Through the looking glass: The role of portals in South Korea's online news media ecology. *Journal of Contemporary Eastern Asia, 18*(2), 16–32. doi:10.17477/jcea.2019.18.2.016

Guzman, A., & Lewis, S. (2020). Artificial intelligence and communication. *New Media & Society, 22*(1), 70–86. doi:10.1177/1461444819858691

Helberger, N., Karppinen, K., & D'Acunto, L. (2018). Exposure diversity as a design principle for recommender systems. *Information, Communication & Society, 21*(2), 191–207. doi:10.1080/1369118X.2016.1271900

Just, N., & Latzer, M. (2020). Governance by algorithms: Reality construction by algorithmic selection on the Internet. *Media Culture & Society, 39*(2), 238–258.

Kaminski, M. E. (2019). The right to explanation, explained. *Berkeley Technology Law Journal, 34*, 189. Retrieved from https://scholar.law.colorado.edu/faculty-articles/1227

Karimi, M., Jannach, D., & Jugovac, M. (2018). News recommender systems. *Information Processing & Management, 54*, 1203–1227. doi:10.1016/j.ipm.2018.04.008

Kemper, J., & Kolkman, D. (2019). Transparent to whom? No algorithmic accountability without a critical audience. *Information, Communication & Society, 22*(14), 2081–2096.

Kim, K., & Moon, S. (2021). When algorithmic transparency failed: Controversies over algorithm-driven content curation in the South Korean digital environment. *American Behavioral Scientist, 65*(6), 847–862. doi:10.1177/0002764221989783

Kitchin, R. (2016). Thinking critically about and researching algorithms. *Information Communication and Society, 20*(1), 1–16.

Kotenidis, E., & Veglis, A. (2021). Algorithmic journalism: Current applications and future perspectives. *Journalism and Media, 2*, 244–257. https://doi.org/10.3390/journalmedia2020014

Kwak, K., Lee, S., & Lee, S. (2021). News and user characteristics used by personalized algorithms. *Technological Forecasting and Social Change, 171*, 120940, https://doi.org/10.1016/j.techfore.2021.120940.

Lewis, S., Guzman, A., & Schmidt, T. (2019). Automation, journalism, and human: Machine communication. *Digital Journalism, 7*(4), 409–427.

Moller, J., Trilling, D., Helberger, N., & van Es, B. (2018). Do not blame it on the algorithm. *Information, Communication & Society, 21*(7), 959–977. doi:10.1080/1369118X.2018.1444076

Newman, N., Fletcher, R., Schulz, A., Andi, S., & Nielsen, R. K. (2020). *Reuters institute digital news report 2020*. Reuters Institute for the Study of Journalism.

Shin, D. (2019). Toward fair, accountable, and transparent algorithms: Case studies on algorithm initiatives in Korea and China. *Javnost: The Public, 26*(3), 274–290. http://doi.10.1080/13183222.2019.1589249

Shin, D. (2020). User perceptions of algorithmic decisions in the personalized AI system: Perceptual evaluation of fairness, accountability, transparency, and explainability. *Journal of Broadcasting & Electronic Media, 64*(4), 541–565. https://doi.org/10.1080/08838151.2020.1843357

Shin, D. (2021). The perception of humanness in conversational journalism: An algorithmic information-processing perspective. *New Media & Society.* http://doi:10.1177/1461444821993801

Shin, D., & Park, Y. (2019). Role of fairness, accountability, and transparency in algorithmic affordance. *Computers in Human Behavior, 98,* 277–284. doi:10.1016/j.chb.2019.04.019

Shin, D., Zaid, B., & Biocca, F. (2022). In platforms we trust? Unlocking the black-box of news algorithms through interpretable AI. *Journal of Broadcasting and Electronic Media, 66*(2), 235–256. https://doi.org/10.1080/08838151.202 2.2057984

Smets, A., Ballon, P., & Walravens, N. (2021). Mediated by code: Unpacking algorithmic curation of urban experience. *Media and Communication, 9*(4), 250–259.

Thurman, N., Lewis, S. C., & Kunert, J. (2019). Algorithms, automation, and news. *Digital Journalism, 7*(8), 980–992. doi:10.1080/21670811.2019.1685395

Wölker, A., & Powell, T. (2021). Algorithms in the newsroom? *Journalism, 22*(1), 86–103. https://doi.org/10.1177/1464884918757072

Zamith, R., & Haim, M. (2020). Algorithmic actants in practice, theory, and method. *Media and Communication, 8*(3), 1–4. https://doi.org/10.17645/mac.v8i3.3395

CHAPTER 8

Human-Centered AI

THE HUMAN-CENTERED AI APPROACH involves users throughout the algorithm development and testing processes, providing an effective experience between humans and AI. Human-centered AI is the system that continuously advances user interaction while offering effective interaction between AI and humans. The need for a human-centered framework for AI has emerged to address the ethical, practical, and legal issues of AI and make it sustainable so that it augments, empowers, and enriches human experiences instead of substituting human capacity. The framework could lead to a fairer, more transparent, accountable, and explainable AI that supports human values, preserves human rights, and promotes user control to steer future AI in the right direction. Important questions include how algorithms fit within a social context, how they can enable meaningful control, and how users can manage algorithm systems effectively. The answers to these questions will guide the development of AI systems that afford humans to see, perceive, create, and behave with confidence and trust. Meaningful human control will play a key role in paving the practical way for realizing meaningful human control over algorithms in AI, as well as in developing extended AI by providing theoretical underpinnings of ethical reflection and by paving the practical way for realizing meaningful human control over algorithms in AI. Extended AI can be designed and should be developed in a human-centered and meaningfully controllable way to contribute to a fairer and more transparent design to forge key positive effects with clear accountability.

DOI: 10.1201/b23083-9

8.1 HUMAN-CENTERED AI AND THE IMPORTANCE OF MEANINGFUL HUMAN CONTROL

Despite drastic advances in the field, AI has not yet achieved the level of complexity needed to represent human cognition (Shin et al., 2022). Even when it becomes possible, AI will require human intervention to govern the sociotechnical considerations of leaving decisions up to algorithms. Human-centered AI has garnered attention from academia and industry for making AI more human-like. What makes us human, and what is meant by human-centered? Our emotions, thinking, and relationships are what make us human. Our human-ness is the very thing that allows us to truly take advantage of AI technologies at our control. The basic inputs, assumptions, steps, and outputs of algorithms should be human-like, accessible, and controllable (Araujo, 2018). Xu (2019) emphasized that AI is not just technological; it is also, and more importantly, social and humanistic. He proposed three goals of human-centered AI: (1) technologically reflect the complexity characterized by human emotions; (2) improve human abilities instead of substituting them; and (3) center on the effect of AI on humans (Xu, 2019). These goals are consistent with goals proposed by other scholars (e.g., Guzman & Lewis, 2020; Shin, 2021b), who stated that the goal of developing an AI system is to simulate humans' attitudes, intelligence, and emotions. With the goal of better understanding human cognition, emotion, and behavior, human-centered AI pushes the boundaries of conventional computing to bridge the gap between humans and AI (Shin, 2021b). Human-centered AI requires systems to understand human intelligence to think in a human-like way. This task is, of course, difficult, as human attitudes, behaviors, and intelligence are dynamic, have complex results, and are contextually dependent (Yang et al., 2020).

One might think of future AI as a terminator-like world where there is no need for human input. However, Ben Shneiderman (2021), the author of *Human-Centered AI*, warns that we should not accept the idea that AI can exceed or substitute for any meaningful notion of human wisdom, intelligence, and responsibility. The overexaggerated capacity of humanoid robots may be misguided and archaic. Unlike scientific fiction's overexaggerated prediction, the future of AI requires more and more human intervention. AI does not exist to replace humans but to enhance and augment human capacities and push society to move forward effectively and efficiently (Thurman et al., 2019). No matter how far AI advances, it cannot reach human-level creative problem-solving without making intuitive jumps to be able to infer from unspecified facts. We humans have conscientious control over the creative process that AI systems do not yet have. For this inherent limitation of AI, algorithms augment, not automate, and industry and human judgment are enhanced, not substituted (Xu, 2019). Thus, human-centered AI reserves human control in a way that warrants AI to satisfy human needs while also functioning transparently, producing fair results, and accepting responsibility for the outcome (Diakopoulos & Koliska, 2016). Human-centered AI designs AI as a tool to assist humanity, not to replace it. The rapid development of AI technologies has produced much hype around their benefits and potential to enrich our human capacity. The future of AI has been discussed from a realistic perspective in terms of the most effective way to leverage AI in our society, and the human-centered AI approach has emerged as a possible paradigm. Human-centered AI is not about being against technological determinism or the pure social construction of technology. The idea of technological determinism is that AI can handle things without any human intervention since algorithms are technically free from the bias that is inevitable to humans. Some extreme futurists predict that algorithms can forge pure autonomous decisions efficiently without errors, whereas humans do not always think rationally or logically. The human-centered approach to AI counters the technology deterministic or autonomous approach by proposing that AI can never substitute for the essential capacities of humans, for example, the full potential of human intellect and wisdom. According to Sundar (2020), we should refuse full domination by algorithms and instead take the primary role in requesting our needs, which will complement and enhance human potential. A human-centered approach promotes a more collaborative interaction between humans and AI, one that keeps the health and safety of humans in mind in both the development and deployment of services. Future AI will involve harmonious collaboration between humans and AI systems, in which both automation and human capacity are essential (Renijith et al., 2020).

The principle of human-centered AI is a recognition that increasingly includes humans in the design process of AI. This recognition repeats the key principle of human-computer interaction, which brings the user into the design loop (Bedi & Vashisth, 2014). The human-centered AI framework is designed to reapply human-computer interaction in a broader AI context. Unlike human-computer interaction, human-centered AI deals with more complicated issues, ethical matters, and contextual aspects, as AI itself is complex. Human-centered AI emphasizes a clearly defined area of ethically loaded contexts within which the system ought to function. This realization is triggered by a growing concern about AI ethics. For example, Google unwillingly admitted, "that incorporating or utilizing AI and machine learning can raise new or exacerbate existing ethical, technological, legal, and other challenges" (Marcus & Davis, 2019, p. 28). Therefore, human-centered AI aims to reduce people's fears of AI threats and increase benefits for society by placing humans at the heart of AI development. Human-centered AI is primarily about placing humans at the center of the system by designing systems that not only support humans but that are also understandable and transparent to humans so that AI becomes explainable and users can consider its suggestions. Including users in the design process means they can monitor for bias in algorithmic decisions. This approach enables a counterbalanced system wherein neither the human nor the machine are entirely autonomous, thus making it easier to identify ways to make the results fairer and more inclusive. The human-centered approach closes the gap between humans and algorithms for the mutual benefit of both.

AI developers have a responsibility to ensure that AI systems operate fairly, transparently, and equitably and that they respect human privacy and serve user needs effectively. More than being data-driven and metric feature-selection approaches, human-centered AI creates deeper and insightful meaning. One way of realizing human-centeredness is to frequently incorporate human users in the design process, especially in the early stages of design. As the user-centered design methods of human-computer interaction are meant for designing interfaces rather than underpinning algorithms (Xu et al., 2019), we, therefore, lack well-established ways of incorporating lay users into the process of designing algorithms in meaningful ways. Shin et al. (2020, 2022, Shin, 2021b) emphasized that human-centered AI is to design, develop, and deploy systems that collaborate with humans in a meaningful way. The concept of meaningful human control will be a key feature in the next generation of human-centered AI. The principle of human control is more than the approach of users-in-the-loop or technical oversight; it requires the duty that lies in the design and development processes (Sartori & Theodorou, 2022). AI systems are based on statistical machine learning of which errors from an inevitable part, often with feedback loops that reproduce, reinforce, and solidify human biases, errors, and irrationalities. As a result of the large number of people affected by AI systems, the number of errors in the form of false negatives and false positives, and of people who are impacted by these errors and ingrained bias, will also increase, triggering the need for oversights. This user control becomes even more critical in high-risk AI applications, such as autonomous machines and autonomous weapons, which demand effective management and governance in the design and operation. Thus, human control should be included throughout the design and use of AI practices. Sociotechnical approaches not only enable effective control but also provide guidelines into the existing structure of bias and unfairness where structures, social relationships, politics, and economic issues are entwined. Integrating this insight into a system design will benefit all stakeholders.

The meaningful control idea is consistent with the extended AI principle as a new paradigm of human–AI interaction (Wienrich & Latoschik, 2021). As the next generation of AI, extended AI will empower humans in the design, development, and usage of algorithms so that users can meaningfully control AI by examining, assessing, and understanding the algorithm's reliability, transparency, fairness, and performance (Santoni de & van den Hoven, 2018). Users of AI can and should be able to audit explanatory models and seek to analyze data to understand models that are consistent with their experiences.

Meaningful human control over AI enables us to bridge responsibility gaps and mitigate them by establishing conditions that promote a suitable attribution of responsibility to humans. The principle of meaningful human control is designed to advance the idea of human involvement, or "being in the loop" (Zanzotto, 2019). However, being involved as a passive agent is not enough to be in control of a human-centered AI because one is still unable to affect any of the functions in the AI model that could come to be seen as even more relevant from an ethical viewpoint. In some cases, users cannot simply understand information or how to influence the process if the algorithms are too complex to understand, and one does not have the technological capacity to respond under the circumstances if they are not in the appropriate position to appreciate the real capabilities of the AI system with which they are interacting. In response to these concerns, many models have been proposed to preserve meaningful human control and user responsibility over AI systems. These include explainable AI (Rai, 2020), interpretable AI (Gunning et al., 2019), responsible AI (Diakopoulos, 2016), usable AI (Xu, 2019), and trustworthy AI (Marcus & Davis, 2019). Shin (2022) argued that meaningful user control over AI can be realized by transparentizing the procedures, providing relevant explanations, and making sure that explanations are understandable and interpretable. In line with this argument, Diakopoulos (2016) claimed that accountability is the most important factor in human-centered AI. In terms of responsibility, it has been proposed that to avoid accountability gaps, algorithmic designers should receive suitable training, and programmers should be made aware of their ethical obligations. Holzinger et al. (2022) emphasized the importance of interpretability in AI design. It may be the case that one or more of these factors is necessary to preserve meaningful human control. Thus, research on human-centered AI is needed, starting with conceptualizing meaningful user control. Eventually, the process of making both digital and physical artifacts will become a partnership in which people will be responsible for configuration, goal setting, steering, high-level creativity, curation, and governance, and AI will be required to enhance human capabilities through automation, innovation, low-level detail work, and the ability to design at scale.

8.2 BUILDING HUMAN-CENTERED AI

Designing AI as concerted work between humans and AI can lead to numerous desirable consequences for users, businesses, governments, and society (Kitchin, 2017). Human-centered AI provides personalized user experiences, leading to increased satisfaction among users. Such personalization can only occur when humans' needs, preferences, and behaviors are considered during the design of algorithms (Zheng et al., 2014). The result of satisfying users is more accurate algorithms built from human values. Industry also benefits from being able to make informed decisions that have the potential to bring the best results through employing predictive analytics in health diagnoses and judicial cases. Informed decisions enable AI to offer more predictable choices and a more dependable solution. By keeping a human-centered focus in AI, we can avoid or effectively minimize the negative consequences of being forced to rely on algorithms.

Building human-centered AI involves two interrelated tasks: (1) AI systems need to be able to understand humans and (2) AI systems need to support humans. Understanding humans involves the task of user experience. AI systems should have a full grasp of how users behave and what they need. This user experience task will make AI more effective, useful, and trustworthy. AI systems that receive their instructions and aims from users will be the ideal type of future AI. However, misunderstanding the user's intentions can lead to system failure, often in significant ways. If the user's instructions are biased, imperfect, or implicit, the results could be disastrous because the AI system will fail to perform the desired output.

Humans might leave part of the goal or the method of accomplishing the goal unspecified because they are used to communicating with other people who have common sense (knowledge) of how something works and how to perform unstated things. By contrast, if not instructed, AI systems can fail because they do not have this common sense and do not share basic human knowledge. However, this kind of failure is considered an error by the human, not an error by the AI system. Reinforcement learning is a machine learning method in which the system uses trial-and-error to master which decisions are maximized in a future reward. One possible way to avoid common sense system failures is for AI systems to understand humans' common sense practices and knowledge. This common sense can be any information normally shared among people from the same culture and society and can be anything from declarative common sense (e.g., cars should stop at red traffic signs) to procedural common sense (e.g., customers should wait for their cues for their turn). Numerous studies have been conducted to create knowledge bases of declarative and procedural common sense information. For example, AI systems have been proposed that can learn procedural common sense information by searching blogs written by other people from a particular culture and society (Shin et al., 2022). Affording AI with common sense and deep learning, instead of simply crunching data analysis, will help establish human-centered AI that we can trust and rely on.

Facilitating humans' understanding of AI systems will involve opening the black box of algorithms to let humans understand the internal process of the AI system. The black box feature leads people to question how algorithms analyze human data and produce results. These reflexive inquiries directly shape users' trust and determine their adoption of AI services. Numerous studies have been done to transparentize algorithms since it is difficult to understand how black box algorithms come to a result based on the data provided (Shin et al., 2022). However, it may even be difficult for algorithm professionals to understand algorithms. Thus, what can the AI system do to develop the perception of fairness, ensure the human that the AI system performs to the best of its capability, and establish trust in its performance? The human-centered approach provides answers to human rationalization. Humans are able to explain their decisions and justify their actions. Like human rationalization, AI rationalization creates an explanation comparable to what humans would give. Explanations of algorithm processes can thus promote perceptions of trust, credibility, rapport, and comfort among lay users of AI systems (Shin, 2021c).

Although explainable AI does not reflect what is actually happening in the underlying AI algorithm, and the explanation may not help humans understand the technical fundamentals of AI algorithms, rationalization is useful, valuable, and relevant in establishing user trust in AI.

Owing to the complicated nature of the tasks involved, human-centered AI brings together practitioners and scholars from various disciplines, such as computer science, psychology, social science, humanities, and engineering, to develop hardware and software, analyze the attitudes and behaviors of people when interacting with AI in diverse social contexts, and acquire the required domain knowledge for particular applications. This collaboration can be challenging due to dissimilar disciplinary approaches and goals. However, the common interest in human-centered AI among this wide variety of disciplines is a common denominator for working together and valuing various ways of realizing human-centeredness.

How can organizations use the theoretical principles of human-centered AI to build AI in a human-centered way? First, the human-in-the-loop principle is the key notion in human-centered AI (Shneiderman, 2021). Users should be involved and engaged throughout the collecting, training, testing, and optimizing processes of designing machine learning models. Humans can dictate the training data used to help the machine learning model learn which conditions and features to identify. Humans should also confirm the correctness of the model's recognition and provide feedback to the model when it predicts incorrect results. Humans should perform the key part of a continuous feedback loop with the model. For instance, in an AI news recommendation process, an automatic news algorithm provides an initial best suggestion, or hypothesis, for a given data, which human editors can use to make their own judgments. News algorithms can also verify human judgments before submitting a recommendation. These types of collaborative models enable the human editor and the news algorithms to work harmoniously to increase the accuracy and effectiveness of news recommendations (Tandoc et al., 2020).

Second, human-centered AI highlights the importance of the context in which algorithms are deployed and used for humans (Renijith et al., 2020). It is critical for designers to grasp the context of use and how sense changes over time. Successful AI design depends on the team's ability to determine and reflect on the desired system's consequences and understand the human and contextual factors affecting those consequences. The system must be able to learn when changes in context occur. Currently, we are witnessing that the consequences of algorithms built a few years ago have much broader repercussions on our society: leading to information spread, the diagnosis of cancers, the predicting of political campaigns, and the development of autonomous cars. These developments require us to examine not only our design philosophies for human-centered AI but also its role in society and the ethical and legal implications of its use. AI should adopt a human method of processing content. For example, chatbots and conversational virtual agents should have a real-world interpretation of text, video, audio, and pictures so they can behave less like decontextualized computers and more like real humans (Go & Sundar, 2019). By interpreting information the same way humans do – by understanding wording and recognizing sensitive cultural and environmental contexts – intuitive understanding would allow AI systems to produce more relevant, accurate, and emotional results. Human-centered AI also needs to be able to understand factors such as the context of use and its environment. Understanding context is a dynamic process that cannot be done at a single point in time. The designers of AI systems need to ensure that the capacity exists to continue user research over time by both humans and the AI system, as the system captures and learns from user behavior in situ (Sundar, 2020). It is also important to note that AI is often designed as part of a broader system like the national health system and the legal decision-making system. In understanding context, there is a need to toggle the focus between the immediate users' behaviors and the context in which the user exists. Research is needed on which mechanisms can be used to collect information and sense or infer user intentions from other parts of larger systems.

Third, human-centered AI requires users and designers to be aware of bias from both humans and algorithms to ensure that we do not rely too much on algorithm judgment or human heuristics. No matter how advanced the AI, bias and errors are inevitable. The important thing is to identify errors and correct them quickly through consistent monitoring. Like human beings, AI cannot be free from errors, as algorithms are based on data. AI designers should comply with legal guidelines, such as the Privacy Act and the General Data Privacy Regulation Guidelines, while developing and using AI services in a human-centered way. They should inform consumers of what data designers are doing with data in each instance and obtain explicit consent. Data should not be transferred or sold to other groups. To protect users' data, data designers have to determine where the data goes, where the design goes, what data they are crunching currently, and what might happen to it in the future.

8.3 EXAMPLES AND FRAMEWORKS OF HUMAN-CENTERED AI

Numerous examples of human-centered approaches to AI services shed light on how the frameworks bring human-centered AI to real-world applications.

8.3.1 Removing Bias in AI-Aided Hiring Process

Some global firms have recently been criticized for biased processes relating to race, gender, disabilities, and sexual orientation when hiring new staff. Despite increasing attention to diversity during recruitment, unconsciously transcended human bias can degenerate the fairness of the process. This bias can be minimized with human-centered AI, thus fulfilling the spirit of diversity. Firms can effectively remove bias from recruiting by utilizing only job-related data as grounds for hiring. With this method, recruiting managers and recruiters may also be better positioned to prioritize diversity and develop a more inclusive recruiting process within their companies. In addition, by focusing on and removing terms that may exclude certain groups, AI algorithms can examine job descriptions to avoid racial or gender bias. With human-centered AI in firms, AI helps transform the industry into a more inclusive environment. There will be a value of equality since AI places every human being in an equal position.

8.3.2 AI-Enabled Conversational Advertising System

Today, advertising is designed to converse with people. With AI-enabled conversational advertising, it is feasible to have meaningful, engaging, and personalized conversations between consumers and advertising. AI-based conversational advertising enables digital marketing to broaden its horizons beyond static and video display assets. This advertising enhances direct-to-customer relations to promote new paths to buy without searching for other sites. Human-centered AI also enables call centers to answer customer needs. AI revolutionizes the modern call center by delivering real-time feedback on user response, predictive analytics to decide when intervention is needed, and in-depth call data analysis to tweak call matching and service quality.

8.3.3 Human-Centered AI in Healthcare and Education

Human-centered AI can analyze large amounts of healthcare data in the form of photographs, clinical research tests, and medical claims, and it can identify patterns and insights that humans typically fail to notice. Human-centered AI can bring efficiency to the field of healthcare and thus facilitate interactions among doctors, patients, and healthcare professionals. Human-centered AI principles can also be applied to educational sectors, and AI played a key role in teaching and learning during the Covid pandemic. By tracking students' individualized needs, AI can save teachers' time by enabling them to tailor their courses to fill knowledge gaps or address areas of concern before students fall behind. Human-centered AI can also assist students with physical and psychological disabilities. For example, text summarization AI makes learning easier for people with dyslexia. For students, human-centered AI means creating customized learning platforms. For teachers, human-centered AI helps predict learning outcomes, enabling them to forge personalized content for each student's objectives and aptitudes.

8.3.4 Human-Centered Recommender Systems

Recommender systems employ user data, including inferred data reflective of user behavior, to personalize user experiences (Shin, 2020). In their simplest form, recommender systems are algorithms that suggest relevant items to customers based on product feedback data. With the rise of platforms such as Netflix, Amazon, and YouTube, recommender systems have increasingly appeared in services for humans, including recommending products that could best interest buyers, suggesting appropriate services and matching their preferences. Despite their wide popularity, current recommender systems face issues such as fairness, transparency, relevancy, accuracy, and privacy, which are yet to be solved effectively.

The goal of human-centered recommender systems research is to develop the algorithms and interactions of recommender systems to better serve the interests of humans. To develop human-centered recommender systems, research and practice are needed to determine the features of recommender systems, the experiences of recommender systems' users, and the relationships between them. Put simply, the human-centered recommendation system zooms in on humans. One goal is to better understand users' perceptions, heuristics, needs, and the impact that recommender systems may have on humans. Another goal is to clarify the issues of fairness, transparency, and accountability of the system, as human-centered recommender systems are concerned with fairness and transparency. As recommender systems have a broad impact on individuals in diverse areas, it is important to determine what is fair and transparent from various perspectives. How can transparency and fairness be achieved? Relevant research (e.g., Holzinger et al., 2022) shows that recommendations that provided understandable and explainable grounds as to why a business partner is recommended were critical to the success of engendering trust in the AI system.

Understanding the user's information processing and cognitive models is the first step in building the recommender system in a human-centered way. For the recommender system's behavior to be explainable to the users in the loop, the system also needs to understand the user's understanding processes of the AI system's task and goal models. The human-centered news recommender model embraces related factors that affect satisfaction, which then influence continuance intention. However, how to embed these principles in human–AI interaction is an open question and compelling task (Yang et al., 2020).

Users' perceived value can be an important part of the recommender system. In the technology acceptance literature, usefulness and ease of use have been widely employed as the basis for analyzing end-user acceptance of technology. Previous studies on consumer acceptance and the adoption of recommendation systems have focused intensely on these beliefs (Shin et al., 2020; Zheng et al., 2014). The idea of an algorithm is related to techniques for modifying the behavior of algorithmic agents over time to improve their usefulness to users (Jung et al., 2017). Users consider the acceptance of news recommender systems based on how useful and convenient they are to use (Knijnenburg et al., 2012). Hence, users' perceived value of usefulness and convenience are essential factors for human-centered recommender systems.

A user's overall satisfaction with a system is related to the perceived performance or service quality of a recommender system (Figure 8.1).

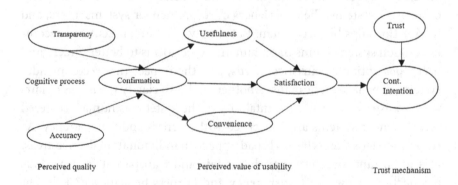

FIGURE 8.1 Human Interation Model of the recommender system.

Shin and Biocca (2018) argued that a user's confirmation level directly influences their satisfaction with technology adoption. Other studies have consistently confirmed a positive correlation between confirmation and satisfaction with AI recommender systems (Karimi et al., 2018; Marcus & Davis, 2019). When users confirm the usefulness of a system, they tend to be satisfied. In the same manner, when users understand the convenience of news recommender systems, their satisfaction levels increase. Transparency and accuracy are determinants of satisfaction (Zheng et al., 2014). Users' intentions, particularly continuing intentions, are determined by their satisfaction with the technology experience. Satisfaction is a psychological effect related to and resulting from a cognitive assessment of the expectation–performance agreement (Renijith et al., 2020). Thus, satisfaction is a key measure of human-centered recommender systems.

Recommendation algorithms and trust metrics comprise the two fundamentals of recommender systems (Figure 8.2; Crain, 2018). Trust plays a facilitating role in technology adoption, particularly in complicated systems (e.g., Zheng et al., 2014). Numerous studies have consistently shown the key role played by trust in the process of evaluation, intention, and diffusion (Xu et al., 2019). Whether users trust certain systems or services affects their assessment, and such assurance influences users' willingness to provide more data to the systems and services (Zheng et al., 2014). In

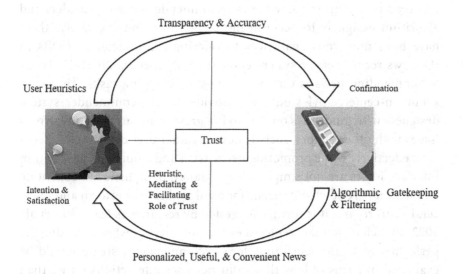

FIGURE 8.2 Feedback loop of trust in algorithm.

the context of news recommender systems, trust is defined as the reliability and accuracy of the news recommendations and the recommender system's capabilities (Marcus & Davis, 2019). Thus, trust signifies how credible and reliable the system is. Many trust factors affect the decision to use technology, but few studies to date have focused on algorithm services, particularly news recommender systems.

The human-centered approach to news recommender systems brings the potential for new service developments and design guidelines. For the developers of news recommender systems or other similar online services, the implications of human-centered AI can help advance the systems' performance and user experience of their products. Human-centered AI recommendations can leverage human experience and qualitatively thick data to capture the deeper needs, aspirations, and drivers that underlie user behaviors in online news. Advanced contextual analytics combine data and user experience to deliver specific behavioral information and produce significantly improved, personalized user experiences. Thus, industries should target users' experiences with algorithms and news recommender systems. Psychological effects and heuristic processes are essential in rationalizing how and why people perceive and understand the issues of news recommender systems and how they use and engage with algorithm-generated news. The main goal of news recommender systems is to help people sort news that is intriguing and relevant to read. Understanding how users search, find, and read news online allows news recommender system providers and algorithm designers to perform more efficiently and effectively. There have been numerous challenges to offering recommended results in the news recommender system context (Bastian et al., 2021). However, recommending news is one of the most challenging tasks. Therefore, a human-centered AI study can provide news recommender system designers with guidelines on how to integrate transparency and fairness issues with other factors, such as how to gather user data/implicit feedback effectively while promoting users' trust and confidence. Algorithm interface issues are not simply esthetic, fancy, or stylistic but constitute an integral part of what it means for an algorithm to perform and interact. In this regard, it is further suggested by researchers (e.g., Shin et al., 2022) to include human-centered evaluation metrics when assessing the performance of an algorithm. News recommender systems should be evaluated in terms of how they influence user interactions rather than by technical measures of performance.

Future news recommender systems need to transcend abstract transparency or numeric accuracy and fulfill actual user needs and perspectives. Thus, understanding user experience is paramount to the usability and success of news recommender systems (Ananny & Crawford, 2018). The human-centered news recommender model will offer insights into how to integrate fairness and transparency with usability factors and behavioral intentions. The goal of news recommender systems is to develop human-centered services. Applying a user-cognitive process to UX design presents users with relevant information. Algorithms that are user-centered and trust-based feedback loops are key to designing such human-centered systems.

8.4 CONCLUSION: ARE AI SYSTEMS INTERPRETABLE, EXPLAINABLE, AND EXPLICABLE?

The key premise of human-centered AI is to consider who will interact with AI instead of designing services only because they are technically feasible. The underlying assumption is that AI systems should be available and communicate in a way that normal nontechnical users can understand and are able to conversant with. The aim of human-centered AI is to develop AI in a way that it can understand how humans think, perceive, communicate, and interact instead of compelling humans to learn how AI systems perform and function. This point gives further two important parameters of human-centered AI: AI systems should (1) be able to understand humans and (2) help humans trust them through fair, transparent, and explainable processes. Within these underlying criteria, relevant research has examined algorithmic experiences, fair and transparent AI, responsible AI, and explainable AI (Rai, 2020). These kinds of AI principles are mechanisms to make autonomous AI systems more sustainable because they will not commit common sense errors, infringe human rights intentionally, or carelessly lead to situations that can lead to harm and conflict.

The design of human-centered AI begins with people and what is desired from the human standpoint by reflecting what people need and what is best for people. Placing humans as the focus and using empathy, equity, and human rights as key values are important (Shin & Biocca, 2018). AI designers should appreciate human-centered AI from human and ethical viewpoints. They should also consider where the data goes, where the design goes, and how to allow users meaningful control of AI, as human control should be embedded through algorithms' operations before, during, and after the use of AI. It will be important to instill user values into algorithms and proactively develop overseeing mechanisms.

Equally important, we need to consider what would be the consequence for humans and what the outcome would be if it were used for negative purposes. Ultimately, when designing and developing AI, we should not believe that AI systems will be perfect. We encourage future efforts in the subject of human-centered AI while safeguarding the human values of the next generation of AI systems.

REFERENCES

Ananny, M., & Crawford, K. (2018). Seeing without knowing: Limitations of the transparency ideal and its application. *New Media and Society, 20*(3), 973–989. doi:10.1177/1461444816676645

Araujo, T. (2018). Living up to the chatbot hype. *Computers in Human Behavior, 85*, 183–189. https://doi.org/10.1016/j.chb.2018.03.051

Bastian, M., Helberger, N., & Makhortykh, M. (2021). Safeguarding the journalistic DNA: Attitudes towards the role of professional values in algorithmic news recommender designs. *Digital Journalism, 9*(6), 835–863. doi:10.1080/21670811.2021.1912622

Bedi, P., & Vashisth, P. (2014). Empowering recommender systems using trust and argumentation. *Information Sciences, 279*, 569–586. doi:10.1016/j.ins.2014.04.012

Crain, M. (2018). The limits of transparency. *New Media & Society, 20*(1), 88–104. https://doi:10.1177/1461444816657096

Diakopoulos, N. (2016). Accountability in algorithmic decision making. *Communications of the ACM, 59*(2), 56–62. https://doi.org/10.1145/2844110

Diakopoulos, N., & Koliska, M. (2016). Algorithmic transparency in the news media. *Digital Journalism, 5*(7), 809–828. doi:10.1080/21670811.2016.1208053

Go, E., & Sundar, S. (2019). Humanizing chatbots. *Computers in Human Behavior, 97*. doi:10.1016/j.chb.2019.01.020

Gunning, D., Stefik, M., Choi, J., et al. (2019). XAI: Explainable artificial intelligence. *Science Robotics, 4*(37), 7120. https://doi:10.1126/scirobotics.aay7120

Guzman, A., & Lewis, S. (2020). Artificial intelligence and communication. *New Media & Society, 22*(1), 70–86. doi:10.1177/1461444819858691

Holzinger, A., Saranti, A., Molnar, C., Biecek, P., & Samek, W. (2022). Explainable AI methods – A brief overview. In A. Holzinger, R. Goebel, R. Fong, T. Moon, K. R. Müller, & W. Samek (Eds.), *xxAI – Beyond explainable AI. xxAI 2020. Lecture Notes in Computer Science*, vol 13200. Cham: Springer. https://doi.org/10.1007/978-3-031-04083-2_2

Jung, J., Song, H., Kim, Y., & Oh, S. (2017). Intrusion of software robots into journalism. *Computers in Human Behavior, 71*, 291–298.

Karimi, M., Jannach, D., & Jugovac, M. (2018). News recommender systems. *Information Processing & Management, 54*, 1203–1227. doi:10.1016/j.ipm.2018.04.008

Kitchin, R. (2017). Thinking critically about and researching algorithms. *Information, Communication & Society, 20*(1), 14–29. doi:10.1080/13691 18X.2016.1154087

Knijnenburg, B., Willemsen, M., Gantner, Z., Soncu, H., & Newell, C. (2012). Explaining the user experience of recommender systems. *User Modeling and User-Adapted Interaction, 22,* 441–504. doi:10.1007/ s11257-011-9118-4

Marcus, G., & Davis, E. (2019). *Rebooting AI: Building Artificial Intelligence We Can Trust.* New York, NY: Penguin.

Rai, A. (2020). Explainable AI. *Journal of the Academy of Marketing Science, 48,* 137–141. https://doi.org/10.1007/s11747-019-00710-5

Renijith, S., Sreekumar, A., & Jathavedan, M. (2020). An extensive study on the evolution of context-aware personalized travel recommender systems. *Information Processing & Management, 57*(1), 102078. https://doi.org/10.1016/j.ipm.2019.102078

Santoni de, S., & van den Hoven, J. (2018). Meaningful human control over autonomous systems: A philosophical account. *Frontiers in Robotics and AI, 5,* 15. doi:10.3389/frobt.2018.00015

Shin, D. (2020). How do users interact with algorithm recommender systems? *Computers in Human Behavior, 109,* 1–10. https://doi.org/10.1016/j.chb.2020.106344

Shin, D. (2021a). The effects of explainability and causability on perception, trust, and acceptance. *International Journal of Human-Computer Studies, 146,* 102551. https://doi.org/10.1016/j.ijhcs.2020.102551

Shin, D. (2021b). The perception of humanness in conversational journalism: An algorithmic information-processing perspective. *New Media & Society.* http://doi:10.1177/1461444821993801

Shin, D. (2021c). Embodying algorithms, enactive AI, and the extended cognition: You can see as much as you know about algorithm. *Journal of Information Science.* http://doi:10.1177/0165551520985495

Shin, D. (2022). How do people judge the credibility of algorithmic sources? *AI and Society, 37,* 81–96. https://doi.org/10.1007/s00146-021-01158-4

Shin, D., & Biocca, F. (2018). Exploring immersive experience in journalism what makes people empathize with and embody immersive journalism? *New Media and Society, 20*(8), 2800–2823. doi:10.1177/1461444817733133

Shin, D., Lim, J., Ahmad, N., & Ibarahim, M. (2022). Understanding user sensemaking in fairness and transparency in algorithms: Algorithmic sensemaking in over-the-top platform. *AI & Society.* https://doi.org/10.1007/ s00146-022-01525-9

Shin, D., Zhong, B., & Biocca, F. (2020). Beyond user experience. *International Journal of Information Management, 52,* 1–11. doi.org/10.1016/j.ijinfomgt.2019.102061

Shneiderman, B. (2021). *Human-Centered AI.* Cambridge, UK: Oxford University Press.

Sundar, S. (2020). Rise of machine agency. *Journal of Computer-Mediated Communication, 25*(1), 74–88. doi:10.1093/jcmc/zmz026

Tandoc, E., Yao, L., & Wu, S. (2020). Man vs. machine? *Digital Journalism, 8*(4), 548–562. https://doi:10.1080/21670811.2020.1762102

Thurman, N., Lewis, S. C., & Kunert, J. (2019). Algorithms, automation, and news. *Digital Journalism, 7*(8), 980–992. doi:10.1080/21670811.2019.1685395

Wienrich, C., & Latoschik, M. E. (2021). eXtended Artificial Intelligence: New prospects of human-AI interaction research. *Frontiers in Virtual Reality, 2*, 686783. doi:10.3389/frvir.2021.686783

Xu, W. (2019). Toward human-centered AI: A perspective from human-computer interaction. *Interactions, 26*, 42–49.

Xu, W., Furie, D., Mahabhaleshwar, M., Suresh, B., & Chouhan, H. (2019). Applications of an interaction, process, integration, and intelligence design approach for ergonomics solutions. *Ergonomics, 62*(7), 954–980. https://doi.org/10.1080/00140139.2019.1588996

Yang, Q., Steinfeld, A., Rosé, C., & Zimmerman, J. (2020). Re-examining whether, why, and how human-AI interaction is uniquely difficult to design. *Proceedings of the 2020 CHI Conference on Human Factors in Computing Systems*. ACM, New York, 1–13. https://doi.org/10.1145/3313831.3376301

Zanzotto, M. F. (2019). Viewpoint: Human-in-the-loop artificial intelligence. *Journal of Artificial Intelligence Research, 64*(1), 243–252. https://doi.org/10.1613/jair.1.11345

Zheng, L., Yang, F., & Li, T. (2014). Modeling and broadening temporal user interest in personalized news recommendation. *Expert Systems with Applications, 47*(7), 3168–3177. doi:10.1016/j.eswa.2013.11.020

Epilogue

THIS BOOK PROJECT WAS initiated and inspired purely by my personal curiosity. While I enjoyed using the convenient and useful personalized recommendations exquisitely generated from AI, I started to wonder whether AI truly knows me and to what extent the algorithm-based recommendations reflect my preferences correctly. Do or can we really know to what extent algorithms are prevalent in our lives? Why are AI opaque, inscrutable, prone to bias, and unaccountable? My curiosities shaped my tasks and eventually my purpose as I teamed up with one of the platform providers that develop AI and algorithms. The task of the research team of this platform was to develop more powerful and effective predictive and prescriptive analytics for users. Soon, we realized that without users' cooperation, consent, and willingness to work with AI, the task is impossible to complete. AI is not a panacea. AI rely on humans to establish parameters and code all predictive decisions. Humans are both resources and targets for AI systems. We shifted our focus from algorithms and machine learning to AI users to propose the next generation of AI, which we term "extended AI." We confirmed that human trust is a key underlying element of AI. We concluded that the way to gain trust is to earn it by building and deploying fair, transparent, explainable, and privacy-preserving AI models. Humans and algorithms interact with one another and are codependent. Human–AI relationships are becoming increasingly important. It is critical to consider all cognitive, ethical, cultural, and legal issues that should be addressed for AI to be considered fully capable of supporting humans in real life.

Unlike the prevalent futuristic vision of AI, human–AI interactions, too, have problems, such as information asymmetry, lack of transparency as to how algorithmic results are curated, absence of mutually constructed interactions, and the possibility of manipulations and

DOI: 10.1201/b23083-10

distortions, particularly in media platforms exemplified as fake news and misinformation. I started to investigate these issues conceptually, theoretically, methodologically, and empirically. Each chapter is the result of a specific research project over the last few years. All the topics are connected, converging into a key question: How can we design and develop human-centered AI?

This book project is designed to contribute to human-centered AI systems capable of performing valuable tasks and being well accepted by humans. A key aspect of this endeavor is enabling AI systems to predict and prescribe relevant user properties and personalize the interaction accordingly in a manner that maximizes both task performance and user satisfaction, abiding by the principles of trust, transparency, interpretability, fairness, responsibility, and meaningful user control. I hope to achieve the initial goal of this book project, which was to contribute to the development of ethical AI systems with which humans can trust and collaborate. Through this book, I offer a critical analysis of the logic and social implications of algorithmic processes. Reporting from the processes of scientific research, the results can be useful and constructive for understanding the interaction and relationship between algorithms and humans. I believe the relationship is and will be an important topic of debate regarding what is at stake while industry and government use AI to reshape the world.

The set of conceptual ideas and design guidelines can serve as a resource for researchers who are aiming at the further development of human–AI interaction theories and practitioners who are working on the design of applications and services that utilize AI technologies. In this book, I have attempted to conceptualize the underpinning principles for human–AI interaction in reference to human–computer interaction. Nevertheless, I must admit that more study and follow-up research are imperative in light of rapid advances in AI and for the clear operationalization of human-centered AI. Time has passed quite fast since I started working on this book project, and AI, related algorithms, and machine learning have radically and continuously changed our lives all this while. I should acknowledge that some of the new research, noble concepts, and new findings in AI may have been overlooked due to fast-changing AI technologies. Based on the findings of this book, I put forward the future research agenda for human-AI interaction.

Index

Printed in the United States
by Baker & Taylor Publisher Services